Humans and Autonomous Vehicles

I0130621

This book provides an introduction to the Human Centred Design of autonomous vehicles for professionals and students.

While rapid progress is being made in the field of autonomous road vehicles the majority of actions and the research address the technical challenges, with little attention to the physical, perceptual, cognitive and emotional needs of humans. This book fills a gap in the knowledge by providing an easily understandable introduction to the needs and desires of people in relation to autonomous vehicles. The book is "human centred design" led, adding an important human perspective to the primarily technology-driven debates about autonomous vehicles. It combines knowledge from fields ranging from linguistics to electrical engineering to provide a holistic, multidisciplinary overview of the issues affecting the interactions between autonomous vehicles and people. It emphasises the constraints and requirements that a human centred perspective necessitates, giving balanced information about the potential conflicts between technical and human factors. The book provides a helpful introduction to the field of design ethics, to enhance the reader's awareness and understanding of the multiple ethical issues involved in autonomous vehicle design.

Written as an accessible guide for design practitioners and students, this will be a key read for those interested in the psychological, sociological and ethical factors involved in automotive design, human centred design, industrial design and technology.

Joseph Giacomin is a professor of Human Centred Design at Brunel University London, UK, where he performs research leading to products, systems and services that are physically, perceptually, cognitively and emotionally intuitive. He has worked for both the American military and the European automobile industry. He has produced more than 120 publications including the books *Automotive Human Centred Design Methods* and

Thermal – Seeing the World Through 21st Century Eyes. He has been a member of the editorial boards of *Ergonomics* and the *International Journal of Vehicle Noise and Vibration* (IJVNV). He is a Fellow of the Chartered Institute of Ergonomics and Human Factors (CIEHF), a Fellow of the Royal Society for the encouragement of Arts, Manufactures and Commerce (RSA), a member of the Associazione Per Il Disegno Industriale (ADI) and a member of the Royal Photographic Society (RPS).

Humans and Autonomous Vehicles

JOSEPH GIACOMIN

Routledge
Taylor & Francis Group

LONDON AND NEW YORK

First published 2023
by Routledge
4 Park Square, Milton Park, Abingdon, Oxon OX14 4RN

and by Routledge
605 Third Avenue, New York, NY 10158

Routledge is an imprint of the Taylor & Francis Group, an informa business

© 2023 Joseph Giacomin

The right of Joseph Giacomin to be identified as author of this work has been asserted in accordance with sections 77 and 78 of the Copyright, Designs and Patents Act 1988.

All rights reserved. No part of this book may be reprinted or reproduced or utilised in any form or by any electronic, mechanical, or other means, now known or hereafter invented, including photocopying and recording, or in any information storage or retrieval system, without permission in writing from the publishers.

Trademark notice: Product or corporate names may be trademarks or registered trademarks, and are used only for identification and explanation without intent to infringe.

British Library Cataloguing-in-Publication Data
A catalogue record for this book is available from the British Library

Library of Congress Cataloging-in-Publication Data
Names: Giacomin, Joseph, author.
Title: Humans and autonomous vehicles / Joseph Giacomin.
Description: New York, NY : Routledge, 2023. |
Includes bibliographical references and index.
Identifiers: LCCN 2022024231 (print) | LCCN 2022024232 (ebook) |
ISBN 9781032334653 (hardback) | ISBN 9781032334646 (paperback) |
ISBN 9781003319740 (ebook)
Subjects: LCSH: Automobiles–Design and construction. |
Human engineering. | Automated vehicles–Moral and ethical aspects.
Classification: LCC TL240 .G53 2023 (print) | LCC TL240 (ebook) |
DDC 629.2/31–dc23/eng/20220907
LC record available at https://lccn.loc.gov/2022024231
LC ebook record available at https://lccn.loc.gov/2022024232

ISBN: 978-1-032-33465-3 (hbk)
ISBN: 978-1-032-33464-6 (pbk)
ISBN: 978-1-003-31974-0 (ebk)

DOI: 10.4324/9781003319740

Typeset in Univers
by Newgen Publishing UK

To Davide, may all his robots prove friendly …

Contents

Figures

Tables

Chapter 1

Introduction

In *I, Robot* (Asimov 1940) Dr Calvin says to her interviewer,"Then you don't remember a world without robots. There was a time when humanity faced the universe alone and without a friend. Now he has creatures to help him; stronger creatures than himself, more faithful, more useful, and absolutely devoted to him. Mankind is no longer alone. Have you ever thought of it that way?"

Not long from now people may find themselves in the similar situation of not remembering what it was like to drive an automobile and of not realising how much they depend upon autonomous road vehicles. Someday soon the images of humans driving motor vehicles might be found only in archives, museums or the history channel. People have the tendency of assuming that those things which existed in the world when they were a child had always existed, and that those things are likely to continue on well beyond their own short lifespan. But change can happen quickly. And in fact we are already beginning to see the first faint and shadowy glimpses of our robotic future on the horizon. The age of full automation is nearly upon us.

Autonomous pods, taxi services, goods delivery robots and other robots are arriving. There are many routine daily activities which most people would be happy to delegate to automated assistants, freeing time and money for other purposes. Many tiring physical activities, in particular, will not be missed. All through human history there has been an ongoing search for labour-saving devices which can help make life more comfortable, and autonomous road vehicles will simply be one further step in a process which is as old as humanity itself.

A brief look back over our shoulders suggests that the arrival of the motor vehicle at the end of the 19th century produced a step change in human mobility. And in the 20th century the motor vehicle went on to

DOI: 10.4324/9781003319740-1

affect nearly every aspect of human life and culture. From the personal sensory novelty of speed to the societal impacts of roads on towns and cities, nearly every aspect of human life was touched in some way. Twentieth century motor vehicles satisfied many needs ranging from the physiological to the self-actualising along the spectrum so elegantly articulated by Maslow (1943).

The change was of course not instantaneous. From the distance of the 21st century and the well-established metaphor of the "automobile" it is easy to miss the fact that things were not always so. Writing in *The Yale Law Journal* in 1908 the legal scholar H.B. Brown noted the criticisms of the motor vehicles of the day including their noise, excessive speed which frightened horses and the running down of pedestrians. Perhaps most problematically, Brown suggested that homes and properties located along thoroughfares dropped in value when vehicular traffic arrived. Despite the virtues, the abrupt introduction of the new technology stimulated sentiments which Brown summarised as "To nearly everyone but the occupants they were an inconvenience; to many a nuisance, and to some a veritable terror".

Historical records document episodes involving angry mobs attacking early motor vehicles (Pearl 2020) in a manner reminiscent of some recent incidents involving autonomous road vehicles (Wong 2018). The public acceptance and legal regulation of motor vehicles took more than three decades to achieve, producing metaphors such as "automobile", "taxi" and "lorry" which we still use today. The introduction of a new technology, lifestyle or value is always complicated, thus consideration of the possible societal reactions is beneficial if difficulties are to be avoided.

As in 1908 when the motor vehicle burst forth from specialist workshops to become an everyday reality, people today are experiencing another revolution in mobility. The recent growth in computing power and in the knowledge of the human brain has produced expert systems, genetic algorithms, neural networks, deep learning systems and other tools for finding patterns in data and for making decisions based on those patterns. In vehicular applications the new algorithms have in turn stimulated the development of new physical sensors, better data formats and faster and more efficient data communication protocols. A new form of mobility and a whole new industry is currently entering the world, that which people are calling "self-driving cars" or "autonomous vehicles".

The starting point of this book is the stereotypical "automobile" in its personal transport or family transport form. While many of the matters which are discussed in this book also apply to other forms of transport such as pods, busses, lorries or trains, it has nevertheless been the intention of the author to narrow the discussion by referring back whenever possible to a form of transport which most people will have experienced themselves, and which many will have come to rely on in deep and meaningful ways. The book does not, however, review current cars or the history of "car culture". Such matters are accurately and engagingly described elsewhere (see for example Macey and Wardle 2009; Barral and Seclier 2017; Gossling 2017 and Meadows 2018).

This book also does not discuss the advanced driver assistance systems (ADAS) which are available on many human-driven automobiles or those which are known to be currently under development. The electronic technologies which are classified as ADAS provide new design challenges and raise important safety issues, but do not fully alter the relationship between the human and the machine. There is still a driver, and, ultimately, the driver is still morally and legally responsible for the driving.

This book also does not discuss the technical challenges involved in developing autonomous road vehicles which can drive safely in rain or snow, communicate efficiently with people from different countries, and provide a range of capabilities and affordances (Herrmann et al. 2018). This book adopts the premise that the technical hurdles will be overcome. It takes for granted that the technological substrate of our new robotic partners will function, eventually. There are simply too many talented individuals already working in the field and too much money already invested for autonomous road vehicles to fail.

And this book also does not discuss the legal issues such as insurance or product liability which are of course essential if autonomous road vehicles are to find their way onto public roads. Despite their importance, such matters are treated comprehensively elsewhere (see for example Channon et al. 2019 and Turner 2018). This book assumes that changes to The Automated and Electric Vehicles Act 2018 (Marson et al. 2020) or the coming into law of superseding legislation will provide appropriate frameworks of insurance status and product liability.

And this book also does not focus on traditional human factors requirements despite their importance in automotive design. Such matters are treated comprehensively elsewhere (see for example Peacock and Karwowski 1993; Schaie and Pietrucha 2000; Bhise 2012 and Akamatsu 2019). It is the author's view that human factors science has progressed tremendously over the last fifty years and that it is now well embedded in the design of automobiles and of other mobility systems via international standards, company standards and commercial best practice.

And, finally, this book does not propose specific autonomous road vehicle designs. While the author does have images in mind of what is doable and what is desirable, those thoughts are not the subject of this book. The autonomous road vehicle meanings and metaphors which are mentioned at various points in this book are mostly incremental developments which are easily imagined and thus helpful for illustrating how automation will change the design focus with respect to traditional human-driven road vehicles.

The focus of this book is on a series of considerations which are widely applicable to humans when they interact with autonomous road vehicles. The point of view is that of the people, rather than that of the technology. This book provides human-focussed information of general relevance, and notes areas where the design of the autonomous road vehicles will prove to be a very different endeavour from that of traditional human-driven road vehicles.

This book focusses on what the new robotic partners might be like from a psychological or sociological perspective and on what their characteristics and capabilities might mean for the people involved. Anthropomorphism, names, meanings, metaphors, interactions and ethics are all discussed in some detail so as to identify as many as possible of the challenges which designers are facing. Autonomous road vehicles are likely to raise as many issues of public acceptance, or more, as their primitive predecessors of 1908. Designers are thus warned that focussing on the technological substrate to the detriment of the psychological and sociological characteristics may be risky.

This book adopts a design point-of-view throughout, thus most of the terms will be familiar to product, system or service designers. The approach is in fact specifically and unashamedly "Human Centred Design" (Giacomin 2014; Giacomin 2017; Gkatzidou et al. 2021) in its widest and most aspirational form. The intention is to place humans at the centre of the autonomous road vehicle project and from that viewing point look around to note the issues which are likely to shape the thoughts and interactions.

Several logical arguments can be proposed in support of a human centred approach but perhaps the simplest is to note that the fixed constraint of the design brief, insomuch as one exists, is the human. Technological capabilities are advancing at a breath-taking pace but humans are changing instead at a slower, more evolutionary, pace. Human physical capabilities, sensory capabilities, psychology, behaviour and societal norms are not static, but do not change as quickly as the technologies are currently changing. As Protagoras had famously suggested (Bostok 1988) more than two millennia ago, "man is the measure of all things".

Chapter 2 of this book introduces the concept of autonomous road vehicle. It summarises a few key points in the history of autonomous road vehicles and provides a few general definitions which are adopted throughout the book. The chapter does not delve in detail into the technical substrate which makes autonomy possible, but instead discusses characteristics of autonomous mobility which affect the human experience.

Chapter 3 of this book introduces the philosophy, or ideology if you prefer, of Human Centred Design. It summarises key concerns and notes how the approach inverts to some degree the temporal order of events which characterise traditional technology-push design processes. The material of the chapter sets the scene by introducing concerns which emerge when an artefact is considered mainly from the point of view of the human.

Chapter 4 discusses the topic of anthropomorphism, i.e. the human tendency towards finding causalities and intentionalities in natural processes and in inanimate objects. It summarises the phenomenon and discusses how the natural tendency will prove more important in the design of 21st century autonomous road vehicles than it did for the motor vehicles of the 20th century.

Chapter 5 discusses naming, an issue which is often assigned little priority by technologists, but which can have profound effects on the interactions which occur between humans and artefacts. The chapter provides examples of the attribute transferral which often occurs with named artefacts, highlighting the possible psychological and sociological implications of name selection.

Chapter 6 discusses the meaning which the autonomous road vehicle or its mobility service may provide. While frequently referred to in design conversations, the word "meaning" is often used inappropriately. The chapter provides an overview of the main forms of meaning and provides examples of how the selected meaning leads directly to requirements and characteristics for the autonomous road vehicle.

Chapter 7 discusses the metaphor which the road vehicle or its mobility service provides. Metaphor is the understanding of one thing in terms of another. It is a way of comparing objects, people or ideas, with the comparison usually being between something simple or familiar on the one hand and something more complex or unfamiliar on the other. The chapter introduces the role of metaphor in design and discusses how a focal metaphor provides a rallying point and a reference against which all characteristics and behaviours can be judged.

Chapter 8 discusses interaction design, interaction design measurements and the interactions which may occur between people and autonomous road vehicles. The complexity of 21st century autonomous road vehicles, particularly their salient characteristic of "autonomy", leads to a rich portfolio of interactions which are of psychological or sociological relevance. The chapter introduces traditional interaction design tools and discusses new developments which appear to be needed for use with autonomous road vehicles.

Chapter 9 introduces several well-known moral theories and notes the dichotomies of "intention vs action" and "individual vs societal" which help to differentiate them. The point is made that none of the moral theories has been found to be universally applicable or universally preferable, and that a well-defined and exceptionless moral theory for universal use does not yet exist. In its place, a series of frameworks, guidelines and standards are presented which provide some ethical guidance of relevance to autonomous road vehicles.

Chapter 10 concludes the book by summarising its main contents and by noting the need for new tools, guidelines and processes to design the complex multifunction machines of tomorrow. In particular, as our ability to understand what they are doing diminishes, trust will become the basis for most of our interactions with them. Tools, guidelines and processes are thus needed to design the trust in. And the moral and legal status of the autonomous road vehicles as either artefact, slave or partner will dictate new functional and ethical constraints which designers will inevitably have to adhere to.

Breaking from traditional automotive semantics, the term "friendly neighbourhood robot" is often used in this book to denote the autonomous road vehicle because robots they will be, with all the imaginable complexity and sophistication. Soon, road vehicles will be performing actions and making decisions which were previously the exclusive domain of humans. Soon, road vehicles will be viewed more as partners than as machines. And, soon, mobility will involve interactions which are more social in nature than they are physical or cognitive.

Humans are profoundly psychological beings. Thus the chapters of this book focus attention on important elements of the human experience when interacting with the friendly neighbourhood robots. The goal is to note how people might interact with the friendly neighbourhood robots and what they might think about them. Key factors which are likely to affect human thoughts, intentions and behaviours are noted, and some of the more obvious deviations from traditional automotive design practice are highlighted. The material is intended to introduce, and hopefully demystify, a few matters which are among the most human.

While not saying all that could be said, the chapters of this book will hopefully further the reader's understanding of the human-facing characteristics of the friendly neighbourhood robots. It is frequently said that design should be at the service of people and that new technologies should be for the benefit of society. It is therefore hoped that the material which is contained in this book can help to fill a current gap in the discussions about automation, and that the ideas which are expressed can highlight what can be done to make our new robotic partners more intuitive, more acceptable and better integrated into society.

The writing of this book has involved a sustained effort to keep the information at an introductory level. The material will therefore hopefully prove accessible to a wide readership despite the complexity of some of the psychological, linguistic, behavioural and philosophical concepts. It is the wish of the author to be as helpful as possible to as many people as possible.

References

Akamatsu, M. 2019, Handbook Of Automotive Human Factors, CRC Press, Boca Raton, Florida, USA.

Asimov, I. 1940, I, Robot, The Gnome Press, New York, New York, USA.

Bhise, V.D. 2012, Ergonomics In The Automotive Design Process, CRC Press, Boca Raton, Florida, USA.

Bostok, D. 1988, Plato's Theaetetus, Oxford University Press, Oxford, UK.

Barral, X. and Seclier, P. 2017, Autophoto, Fondation Cartier pour l'art contemporain, Paris, France.

Brown, H.B. 1908, Status Of The Automobile, The Yale Law Journal, Vol. 17, No. 4, February, pp. 223–231.

Channon, M., McCormick, L. and Noussia, K. 2019, The Law And Autonomous Vehicles, Informa Law From Routledge, Abingdon, Oxon, UK.

Giacomin, J. 2014, What Is Human Centred Design?, The Design Journal, Vol. 17, No. 4, pp. 606–623.

Giacomin, J. 2017, What Is Design For Meaning?, Journal Of Design, Business & Society, Vol. 3, No. 2, pp. 167–190.

Gkatzidou, V., Giacomin, J. and Skrypchuk, L. 2021, Automotive Human Centred Design Methods, Walter de Gruyter GmbH, Berlin, Germany.

Gossling, S. 2017, The Psychology Of The Car: automobile admiration, attachment, and addiction, Elsevier, Amsterdam, Netherlands.

Herrmann, A., Brenner, W. and Stadler, R. 2018, Autonomous Driving: how the driverless revolution will change the world, Emerald Group Publishing, Bingley, UK.

Macey, S. and Wardle, G. 2009, H-Point: the fundamentals of car design & packaging, Art Center College of Design, Design Studio Press, Culver City, California, USA.

Marson, J., Ferris, K. and Dickinson, J. 2020, The Automated And Electric Vehicles Act 2018 Part 1 and Beyond: a critical review, Statute Law Review, Vol. 41, No. 3, pp. 395–416.

Maslow, A.H. 1943, A Theory of Human Motivation, Psychological Review, Vol. 50, No. 4, pp. 370–396.

Meadows, J. 2018, Vehicle Design: aesthetic principles in transportation design, Routledge, New York, New York, USA.

Peacock, B. and Karwowski, W. 1993, Automotive Ergonomics, Taylor & Francis, London, UK.

Pearl, T.H. 2020, Hands Off The Wheel: the role of law in the coming extinction of human-driven vehicles, Harvard Journal Of Law & Technology, Vol. 33, No. 2, pp. 427–475.

Schaie, K.W. and Pietrucha, M. 2000, Mobility And Transportation In The Elderly, Societal Impact on Aging Series, Springer Publishing Company, New York, USA.

Turner, J. 2018, Robot Rules: regulating artificial intelligence, Palgrave Macmillan, Cham, Switzerland.

Wong, J.C. 2018, Rage Against The Machine: self-driving cars attacked by angry Californians, The Guardian, March 6th.

Chapter 2

Autonomous Road Vehicles

History Of Autonomous Road Vehicles

Narratives (Maurer et al. 2016) suggest that the first attempts at removing the human driver from within the road vehicle took place in the 1920s using radio technology. RCA Corporation demonstrated a radio-controlled automobile in Dayton Ohio in 1921. And in 1925 the Houdina Radio Control Company demonstrated a two-car team in New York City consisting of a controller vehicle and a controlled vehicle. Such early experiments were not about autonomous driving in the sense that the automobile drove itself, but nevertheless showed that there were alternatives to the need for a human driver sat at the wheel.

From the 1930s onwards various organisations tested guide-wire systems for automobiles and lorries, with the goal of achieving safety and security during motorway drives. By the 1970s several automobile manufacturers had used video cameras to follow road markings or had developed electronic guidance systems which could follow magnets or transponders in the roadway. Examples were the automobiles developed by Tsukuba University and Toyota from 1977 onwards which used video cameras to follow lane markings at speeds of up to 50 km/h. Automobiles had begun to drive themselves, even if only for short periods and in simple circumstances.

In the United States in the 1980s the Defence Advanced Research Projects Agency (DARPA) funded projects such as Carnegie Mellon University's Navlab and ALV which developed computer-controlled vehicles. At approximately the same time in Europe the Prometheus Project, which ran from 1987 to 1995, developed AI, vehicle-to-vehicle communications, vehicle-to-environment communications and traffic modelling. The widespread availability of powerful and relatively compact

DOI: 10.4324/9781003319740-2

computers was by this point making it possible to perform sophisticated detection and navigation tasks on the road in real time.

Commercial investment began to increase dramatically after DARPA issued its now famous grand challenges. The automotive competitions offered cash prizes and provided media visibility at a time when much progress was being made in robotics and artificial intelligence. The first DARPA Grand Challenge competition was held in 2004 in the Mojave Desert along a 240 km route. None of the vehicles finished the course. The best result was that of the Carnegie Mellon team whose car completed 11.78 km. No winner was declared, but the media coverage was extensive, and when the challenge was repeated in 2005 the route was completed by five teams. The winning robot of 2005, Stanford University's Stanley (Thrun et al. 2006), finished the route in 6 hours, 53 minutes and 58 seconds.

In more recent times the year 2017 saw Waymo's announcement that it was testing driverless cars without a safety driver. Then in 2018 Waymo announced that its vehicles had travelled in automated mode for over 16,000,000 km and that the total was increasing at a rate of 1,600,000 km per month. By December 2018 the company was the first in the world to commercialise a fully autonomous taxi service in Phoenix, Arizona. Driverless cars were coming close to becoming a commercial reality.

Viewed from a distance, it seems that improvements in autonomous driving have been in close step with improvements in the available electronic technologies. And with the growing power and relatively low cost of current computers, it would also seem likely that more driving tasks and responsibilities will be handed over to automation in the coming years. There is currently no obvious sign on the horizon of a major technical impediment to further growth in computational abilities via hardware and via software, thus no obvious reason for doubting that fully autonomous abilities will eventually be reached. As the electronic sensing, pattern recognising and decision making abilities improve, there will inevitably come a point at which the machine abilities will be close enough to those of licensed human drivers to take over responsibility for the driving. Soon, humanity will not be facing the roadway universe alone.

Autonomous Road Vehicles

So what exactly is an autonomous road vehicle? The UK national CCAV (Centre for Connected and Autonomous Vehicles 2020) has defined an automated vehicle to be a "vehicle designed or adapted to be capable, in at least some circumstances or situations, of safely driving itself and may lawfully be used when driving itself, in at least some circumstances or situations, on roads or other public places in Great Britain."

Up to now the concept of automated road vehicle has usually been understood to mean a traditional road vehicle which has been equipped with additional functional capabilities which permit it to drive itself in some situations some of the time. For many people today, perhaps for most, the metaphor which comes to mind and which is visualised when thinking about autonomous road vehicles is something along the lines of the vehicle which is shown in Figure 2.1. It is a stereo-typical automobile which is equipped with additional electronic sensors and processing equipment to enhance the ability to make driving decisions.

More recently, however, there has been a growing realisation that self-driving means more than additional decision making on the part of the vehicle. If the sensors, connectivity, maps, algorithms and actuators can be made safe and reliable, then there is scope for eliminating the human driver altogether and for reconsidering major elements of the vehicle. The elimination of the driver's seat, manual driving controls and human driver leads to a very different type of road vehicle. Form and function can be reconsidered, and departures from traditional automotive stereotypes are possible.

Figure 2.1 A traditional road vehicle equipped with self-driving capabilities.

Source: ©unitysphere/123RF.com

Figure 2.2 A non-traditional road vehicle designed around self-driving capabilities.

Source: ©artzzz/123RF.com

Figure 2.2 presents a typical example of the current departures from traditional automotive stereotypes. The robo-taxi has some features of traditional road vehicles, but exploits the opportunities offered by the elimination of the manual driving controls and by the rethinking of the seating arrangements and human ingress/egress. Several experimental deployments of similar robo-taxis are underway in areas of major infrastructure such as train stations, sports arenas and airports where the routes tend to be short, well defined and in some cases separated from the general traffic circulation by separate lanes.

Whether enhanced versions of existing human-driven automobiles or instead non-traditional road vehicles such as robo-taxis, current autonomous vehicles are equipped with electronic systems which can control the vehicle for periods of time in absence of human intervention. And some autonomous road vehicles are also equipped with dedicated electronic systems for interacting with other road vehicles, with passengers, and with other road users such as pedestrians. Importantly, the electronic systems are more tightly connected, coordinated and integrated than what had been past automotive practice in which individual subsystems such as the engine control, steering, braking and climate control had tended to operate somewhat independently of each other.

While the components and subsystems vary in number and complexity from manufacturer to manufacturer, a logical subdivision of the systems can be suggested based on the information which is processed and on the interactions which occur with other vehicular systems, with external machines and with humans. A list of autonomous road vehicle electronic systems appropriate for use by non-specialists is presented in Table 2.1.

TABLE 2.1 Autonomous road vehicle electronic systems.

Perception System	Environment perception and position localisation. The system can include one or more ultrasonic sensors, optical cameras, infrared cameras, RADAR sensors, LIDAR sensors and global positioning system (GPS) receivers. Algorithms perform data filtering, data fusion and map location detection (see for example Ahangar et al. 2021 and Rosique et al. 2019 for details).
Planning System	Mission planning, behavioural planning and motion planning. The system can include algorithms which integrate the vehicle's global map position with the sensed local distances to obstacles and landmarks, and typically involves algorithms for determining the control actions such as accelerations, decelerations, lane keeping and lane changing (see for example Rosique et al. 2019, Leon and Gavrilescu 2021 and Yoon et al. 2021 for details).
Control System	Trajectory prediction, trajectory tracking and vehicle hardware actuation (steering, brakes, etc.). The control algorithm determines the vehicle dynamic parameters including steering angle, longitudinal acceleration and driving style adjustment, and monitors instantaneous values to maintain them within a tolerance band (see for example Leon and Gavrilescu 2021 and Yoon et al. 2021 for details).
Communication System	Vehicle-to-vehicle (V2V), vehicle-to-infrastructure (V2I) and vehicle-to-everything (V2X) systems. Transmission and reception technologies collect driving-relevant and passenger-relevant data for use in decision making by the vehicle and by the passengers (see for example Ahangar et al. 2021 for details).
Internal HMI System	Human–machine interfaces for use by passengers. The system typically includes vehicle lighting, display screens, loudspeakers and other devices for informing passengers, and buttons, optical cameras, microphones and other devices for accepting requests or commands from the passengers (see for example Ataya et al. 2021 and Xing et al. 2021 for details).
External HMI System	Human–machine interfaces for use with other road users and pedestrians. Typically includes vehicle lighting, display screens, loudspeakers and other devices for informing other road users and/or pedestrians, and optical cameras, microphones and other devices for detecting signals, warnings and requests from other road users and/or pedestrians (see Carmona et al. 2021 for further details).

TABLE 2.2 SAE levels of automation.

Level 0	The automated system issues warnings and may momentarily intervene but has no sustained control of the vehicle.
Level 1	The driver and the automated system share control of the vehicle. Examples include Cruise Control, Adaptive Cruise Control, Parking Assistance, Lane Keeping Assistance and Automatic Emergency Braking.
Level 2	The automated system takes full control of the vehicle and the driver monitors the vehicle and is prepared to intervene.
Level 3	The driver can safely turn their attention away from the driving task. The automation will handle situations that call for an immediate response, but the driver must be prepared to intervene when called upon to do so. Examples include Traffic Jam Chauffeur and Automated Lane Keeping System.
Level 4	Similar to level 3 but requiring no driver attention in limited spatial areas (geofenced) or under special circumstances.
Level 5	No human driving intervention required in any driving content.

In recent years much of the discourse about the design of the autonomous road vehicle subsystems has revolved around the levels of automation defined by the American Society of Automotive Engineers (SAE 2014). The SAE definitions (see Table 2.2) describe the roles and responsibilities of the road vehicle and of the human, in steps from little automation to full driving automation.

The SAE levels of automation provide an engineering description of what a semi-autonomous or fully autonomous road vehicle can do with respect to the nominated human driver. They describe an incrementally increasing list of driving tasks which enter the domain of the road vehicle and which exit the immediate responsibility of the nominated human driver. They provide a roadmap for how the decision making of the automation can be increased, in steps, from that of traditional human-driven systems to robotic partners which deal with nearly every aspect of route planning and route following.

The incremental steps outlined by the SAE levels have stimulated much research about the nature and characteristics of shared responsibility scenarios. SAE levels 1, 2, 3 and 4 all imply the existence of situations in which handovers of control occur from the nominated human driver to the vehicle, or from the vehicle to the nominated human driver. At an early stage in the development of the current forms of partial automation the transitions were noted to be cognitively complex and safety critical. Issues including situation awareness (Endsley et al. 2004), inattentional blindness (Mack and Rock 1998), reaction time (Groeger 2000) and mental after-effects (Stanton et al. 2021) were all observed to affect the handovers of control between the vehicle and the human. Time delays and errors in handover were noted to occur frequently, raising concerns.

A substantial body of information has now emerged (see for example Eriksson and Stanton 2018 or Stanton et al. 2021) which quantifies human take-over time, which describes the effects of the sensory modalities used, and regarding the opportunities for individual personalisation of the handover process. It has been repeatedly noted that take-over requests by the vehicle can be problematic due to the nominated human driver being engaged in other tasks at the time or due to the time delay associated with the shifting of the cognitive scene. And, unlike the situation with the autopilot and other flight automations of aircraft, the road vehicle environment offers much less room, literally, for error. The short physical distances involved in many driving scenarios can lead to safety-critical situations for even small time delays or minor errors of interpretation. In fact, several highly publicised accidents involving road vehicles which were equipped with advanced driver assistance systems (ADAS) have highlighted the dangers of shared control, and have focussed attention on how the errors can be initiated by either the road vehicle element or the human element of the combined system.

The SAE levels of automation provide nomenclatures and definitions which are directly applicable to the engineering aspects of the road vehicle. They are one possible way of categorising the functional characteristics and of discussing them. The levels are incremental in logic and nature, though they probably do not represent equal steps in terms of complexity and cost. It is important to note however that the levels are mostly descriptions of the road vehicle's capabilities rather than of what the human might think or do.

The SAE definitions do not constitute a Human Centred Design approach. And they do not provide a set of evolutionary steps which are approximately equal in terms of their impact on people. The approach does not say much about who might be in the vehicle, what they might be expecting from the vehicle or how they might wish to interact with the vehicle. Defining the engineering characteristics of a control handover is not the same as defining its psychological nature, sociological impact or meaning for the people involved.

The most obvious use of the SAE levels of automation is as an engineering roadmap which bridges the gap between the existing human-driven road vehicle technologies and those technologies which will be present in the fully self-driving and fully autonomous road vehicles of the future. The levels provide descriptions of possible advances in technology which will get us to the point of full autonomy. They help to understand what is needed technologically and provide the nomenclatures and definitions needed for the discourse. They are not, however, proposals of what future mobility may actually be like from the point of view of the people involved.

Autonomous Electrical Road Vehicles

Autonomous road vehicle discourse has usually focussed on what can be referred to as the guidance system. Much has been speculated in relation to how effective the current prototype vehicles have been at avoiding accidents, how closely they resemble humans in their driving, whether indeed it is a good idea to drive like a human, and how the automation handles ethical considerations such as the choice of what to run into, if, in fact, it is not possible to avoid an accident. The ability of automation to replace humans and the moral and legal implications of that change have usually been the stimulants of public debate.

Developments in the automotive sector are however also rapidly modifying that which is guided. The mechanical substrate of the road vehicle is also evolving, expanding the realm of the possibilities. In parallel with the developments in the sensors, databases and algorithms of the guidance systems there is also a rapidly developing reality of the electrification of the road vehicle fleet (Enge et al. 2021). While not strictly a matter of autonomy, the growing electrification of road vehicles is nevertheless changing its substrate, eliminating many traditional components and simplifying many others. The design opportunities offered by the automated driving systems are being multiplied by the many simplifications and repositionings which electrification permits for the underlying componentry of the road vehicle. With respect to 20th century automobiles, the autonomous road vehicles of the 21st century are likely to be as different in their material structure as they will be in their driving control.

Where petrol and diesel internal combustion engines were the dominant power technologies in the 20th century, the 21st century appears to be heading for a generalised adoption of batteries and electric motors. Beyond the reduction in the number of moving parts, which has been estimated to be up to two orders-of-magnitude (Seba 2018), there are also opportunities for consolidation and simplification of battery packs and auxiliaries. Inspection of the mechanical and electrical components which are visible in the examples of Figure 2.3 and Figure 2.4 suggests a greater amount of available space, and fewer physical constraints on its

Figure 2.3 Indicative mechanical layout of current petrol or diesel automobiles.

Source: Henrik5000

Figure 2.4 Indicative mechanical layout of current electric automobiles.

Source: 123RF.com

use, in the case of a road vehicle which moves via batteries and electrical motors.

It thus appears likely that the widespread adoption of both electrical powertrains and self-driving technologies will provide opportunities for eliminating many current components, simplifying others, and offering greater levels of personalisation of the physical packaging of the vehicle. As both the number of the underlying components and their complexity diminish, it should prove possible to offer a greater variety of vehicle platforms while meeting the same manufacturing and cost constraints.

While not strictly a matter of autonomy, the increased spatial and geometric design freedoms should permit greater scope for customisation and personalisation of the vehicle packaging (Macey and Wardle 2009). Presumably, this would in turn provide opportunities for meeting a greater number of mobility needs and for supporting further human desires which involve an element of movement or travel. More so than their petrol or diesel counterparts, autonomous electrical vehicles should greatly facilitate the achievement of new vehicle types, new vehicle uses and new vehicle meanings.

Autonomous Mobility

Dictionary definitions of the word "mobility" suggest concepts such as the movement of individuals or groups from place to place, from job to job, or from one social or economic level to another. The word refers more to what happens to the people than it does to the means of transport.

In its report "The Future Of Mobility" (Vallance and Norman 2019), the UK Government Office for Science stated that "Mobility – the movement of people and goods – is generally not an end in itself. Its value lies in the accessibility it provides and how this contributes to the functioning and quality of people's lives, as individuals and as a society". The report went on to state that "Mobility is essential for social cohesion, widening people's opportunities and improving their health and well-being". The United Nations (2006) has gone as far as declaring mobility to be a basic human right.

The term "autonomous mobility" refers to the use of autonomous road vehicles as opposed to mass transit systems or traditional forms of personal transport such as motorcycles, automobiles or vans. Since the existence of road vehicles which are capable of driving themselves is a relatively new phenomenon, so too is the term "autonomous mobility". At the current point in time the term is mostly associated with small-scale local transport experiments based on autonomous pods or busses which move along short and relatively constrained routes (see Marres

2020 for UK examples). The field of autonomous mobility is however destined to grow as new technologies and new laws come into effect. Autonomous road vehicles will inevitably be added to the mix of transport options which form the physical substrate of mobility, and if proven economical might one day constitute the bulk of many transport fleets.

It has been suggested (Kakihara and Sorensen 2001) that the concept of mobility should be thought of as involving three distinct dimensions: spatial, temporal and contextual. The spatial dimension is the most traditional and familiar consideration, referring to the distances and landscapes which are travelled and to the effects that such travel has on people's social and economic lives. The temporal dimension refers instead to the time consumed by the journeys, and came to prominence in the 20th century as new technologies reduced travel times and dramatically accelerated the pace of life. Future forms of mobility which base themselves on autonomous road vehicles may deviate somewhat from the spatial and temporal characteristics of 20th century human-driven road vehicles, but as many of the physical and infrastructural constraints will remain the same, or at least similar, the deviations may not be large.

Where autonomous mobility may instead deviate more substantially is in the contextual dimension. Designers are well aware of the influence of context on people's emotions, memory recall, thinking and actions. And they routinely design (Holtzblatt et al. 2005) artefacts based on knowledge of the full situation consisting of location, intentions, physical interactions and social interactions. As suggested by Kakihara and Sorensen (2001), "Human action is inherently situated in a particular context that frames and is framed by his or her performance of the action recursively. Such contextuality, or situatedness, of human action is critical for capturing the nature of interaction".

The contextual dimension offered by the friendly neighbourhood robots will almost certainly prove to be different from, and more diverse than, that offered by current automobiles. For example, the elimination of the design constraints which were imposed by the physical presence of the nominated human driver will lead to differences in the journey types, journey distances and levels of privacy provided. The lack of a driver changes both the physical layout and the social environment within the road vehicle. It makes possible different uses of the space and different compositions of the passenger group. Where the contextual dimension of a typical 20th century automobile might have involved the needs of a lone driver or of a small group of colleagues, friends or family members who had to get from one place to another, both the group composition and the activities which the people wish to perform may be different in the future.

And while most current design efforts are focussing on the engineering properties of the guidance system so as to achieve the greatest possible driving safety, the automation will eventually spread more widely once acceptable levels of safety are achieved. The automation will make possible interactions which support human needs or desires which transcend simple travel, via additional activities which are today usually within the realm of work, shopping, health or leisure. The friendly neighbourhood robots will inevitably offer a range of interactions with people across different sensory modalities, bridging space and time in ways that were simply not possible with 20th century automobiles. The contextual dimension of the friendly neighbourhood robots will be different, and is likely to be much richer than what was possible to achieve with the human-driven road vehicles of the past.

Mobility studies have traditionally leveraged scientifically gathered information about the physical characteristics of the journeys and about the psychological motivations behind them. Typical examples of such information are shown in Figures 2.5, 2.6 and 2.7, which illustrate the kinds of facts which can be gathered when attempting to answer questions about who is travelling, where they are going and how they are getting there.

While helpful, information of this type has in recent years begun to reach its limits in terms of the insights which can be achieved. As new mobility technologies have joined the fray, the options for any given journey have multiplied and the range of considerations influencing the travel decisions has expanded. Besides considering the scientifically gathered and statistically analysed data about the movements themselves, businesses and governmental bodies are increasingly basing decisions on additional psychological considerations such as the values, meanings and decision-making strategies involved.

One summary of the factors which can influence mobility behaviour was suggested by Hunecke (ADD HOME 2009) and is reproduced here in adapted form as Figure 2.8. As shown on the diagram the factors involved were subdivided into two groups: personal factors and external factors.

The factors which influence individual mobility behaviour include both personal and external factors which are independent of the design of the autonomous road vehicle. Factors such as the person's demographics, the distance to travel, the natural environment involved and the local public transport policy are all somewhat outside of the control of the autonomous road vehicle designers. While the design process can support choices in these arenas, it would not be expected to determine or even drive those choices.

Average distance per person per day
(kilometres)

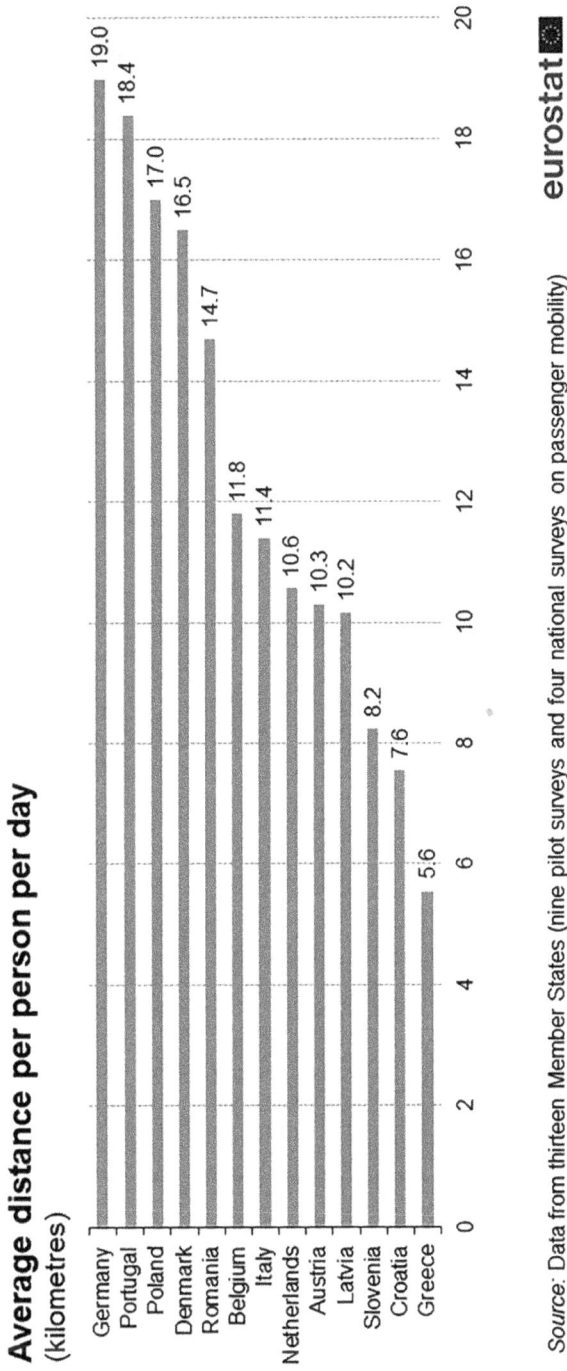

Figure 2.5 Average distance travelled per person per day in 12 EU countries.

Source: Data from thirteen Member States (nine pilot surveys and four national surveys on passenger mobility)

Source: Eurostat 2021

Travel distance per person per day by main travel mode for urban mobility on all days
(%)

	Belgium	Denmark	Germany	Greece	Croatia	Italy	Latvia	Netherlands	Austria	Poland	Portugal	Romania	Slovenia
By car as driver	54.4	53.8	58.0	44.6	59.6	63.7	54.8	49.6	50.6	48.2	57.3	30.4	65.2
By car as passenger	16.3	11.3	11.8	15.4	13.3	10.6	13.0	12.6	13.5	10.6	12.9	26.4	15.4
By taxi (as passenger)	0.1	0.3	0.2	1.3	0.4	0.2	0.5	0.0	1.1	0.0	0.4	2.5	0.2
By van/lorry/tractor/camper	0.0	8.1	2.2	0.8	2.3	0.1	0.0	0.0	0.0	0.0	0.0	0.0	1.1
By motorcycle and moped	0.8	0.9	0.6	7.0	0.1	2.8	0.3	2.0	1.0	0.6	1.3	0.1	0.2
By bus and coach	4.3	4.1	2.3	11.5	9.9	7.2	13.1	3.7	4.0	25.9	10.8	27.6	6.8
Urban rail	2.8	4.4	5.4	12.8	5.0	2.5	4.9	0.0	13.0	2.9	4.0	1.9	0.0
By train (regular and high speed)	8.6	5.5	8.6	0.1	2.8	3.8	5.2	7.5	9.0	2.9	5.1	3.9	1.3
Aviation and waterways	0.0	0.0	0.1	0.0	0.2	0.1	0.0	0.0	0.0	0.0	0.5	0.0	0.0
Cycling	6.6	7.5	5.5	0.5	2.1	1.9	2.2	16.0	3.4	4.7	0.5	0.3	3.3
Walking	3.8	4.1	4.0	5.8	4.5	6.8	6.1	5.1	3.9	1.8	5.8	6.9	6.5

Note: the shares of the main travel modes may not add up to 100% due to the small residual category "Other/unknown", not shown in the table.

Source: Data from thirteen Member States (nine pilot surveys and four national surveys on passenger mobility)

eurostat ◉

Figure 2.6 Average distance travelled per person per day in 12 EU countries subdivided by travel mode.

Source: Eurostat 2021

Distribution of distance travelled per person per day by travel purpose for urban mobility on all days

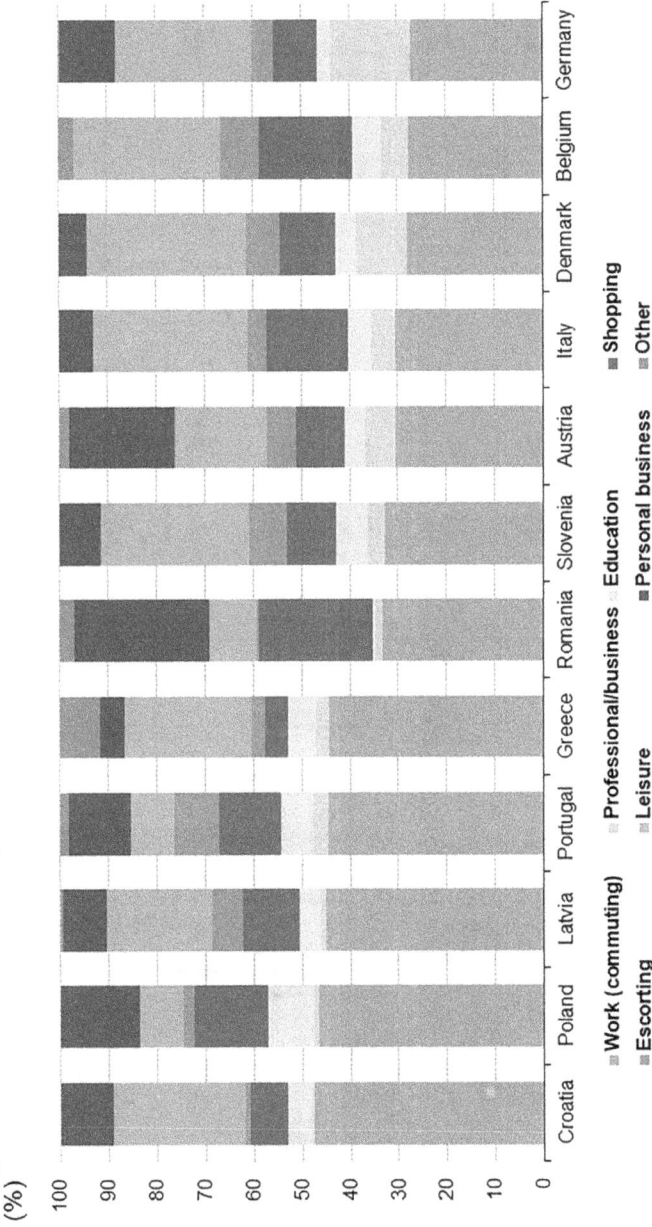

Source: Data from twelve Member States (eight pilot surveys and four national surveys on passenger mobility)

Figure 2.7 Average distance travelled per person per day in 12 EU countries subdivided by travel motivation.

Source: Eurostat 2021

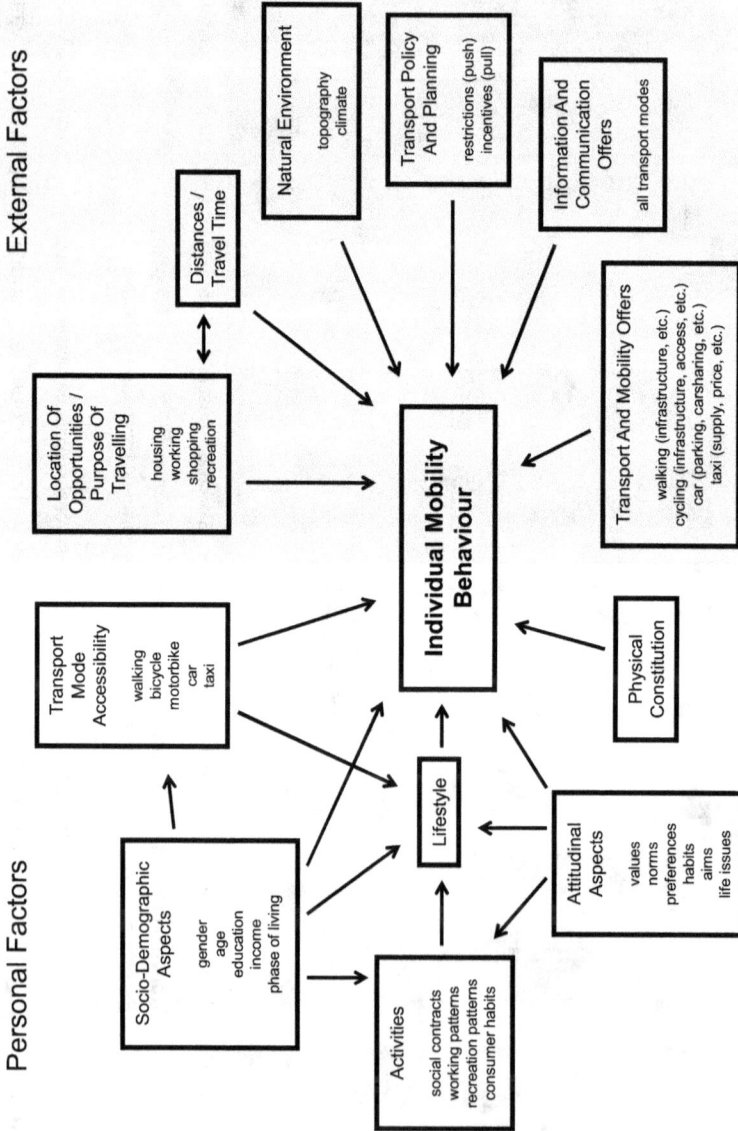

Figure 2.8 Factors which influence individual mobility behaviour.

Source: adapted from ADD HOME 2009

There are, however, also factors which can be directly influenced by appropriate design choices. For example, providing transport information and accommodating the user's physical condition, lifestyle or socio-economic characteristics fall within the remit of the autonomous road vehicle designer. In each case the targeted characteristic or capability can easily become a focal point for the design, branding and marketing of the autonomous road vehicle. The selection of the specific characteristics or capabilities to prioritise would therefore be expected to be an important step in the design of a friendly neighbourhood robot, and a key decision in its associated business model development.

The benefit of such focussing has in fact been among the main findings of workshops (see for example Strömberg et al. 2018) and surveys (see for example Pettigrew et al. 2018) which have investigated public expectations. From the support for disabled individuals and elderly individuals (Harrow et al. 2020) to low-cost mobility-as-a-service (Jittrapirom et al. 2017) solutions, many opportunities have already been identified which involve more specialised forms of mobility than what was considered economically feasible with 20th century road vehicles. New road vehicle metaphors such as the "shopping car" or the "doctor on wheels" seem likely to be on our roads at some point soon.

Figure 2.9 presents a range of concepts which illustrate what people currently think might be possible given the new design freedoms. The elimination of the driving controls, the rethinking of the seating arrangements and the approach chosen for ingress and egress can all be used to focus and specialise the vehicle. If an inexpensive basic autonomous road vehicle platform can be manufactured which can accommodate a variety of specialist bodies, then there will be few limits to where human imagination may lead. And if also proven viable economically, such autonomous road vehicles should help to address several additional societal needs and should help to provide mobility in localities where achieving adequate mobility has traditionally proven elusive.

The deployment of increasingly sophisticated forms of automation and the freedoms offered by the absence of the dedicated human driver will lead to road vehicles whose functions and aesthetics are dictated more by what they can do for people than by what they can carry or where they can go. And as the friendly neighbourhood robots grow in number and specialisation it should become easier to provide mobility options which fit each person individually in the manner of a comfortable favourite glove. Like the motor vehicles of the 20th century, the friendly neighbourhood robots of the 21st century are likely to have a profound impact on human society. Everyday life will never be the same again.

Figure 2.9 Examples of autonomous road vehicle concepts.
Source: Henry Leeson

Conclusions

This chapter has provided a short history of the autonomous road vehicle and has briefly introduced the concepts of autonomous road vehicle,
autonomous electric road vehicle and autonomous mobility. A set of six
subsystems was introduced to assist the general understanding of the
functions which are usually part of an autonomous road vehicle. The
Perception System, Planning System, Control System, Communication
System, Internal HMI System and External HMI System each process specific items of information and control specific interactions with humans.

And the levels of automation defined by the SAE were presented
to assist the understanding of the degrees of control sharing between
human and vehicle which are possible as road vehicles evolve towards
fully self-driving robots. It was noted that the SAE levels provide useful nomenclatures and definitions which are applicable to the engineering aspects of the road vehicle, but do not constitute a Human Centred
Design approach. Defining the engineering characteristics of a control
handover is not the same as defining its psychological nature, sociological impact or meaning for the people involved.

While not strictly a matter of autonomy, the trend of road vehicle
electrification was also mentioned. Adopting electric power as the
basis for the road vehicle leads to a dramatic reduction in the number

of components and to their simplification and repositioning around the vehicle. This provides the designers greater freedom for pursing ingress/ egress, seating and internal packaging arrangements which were not previously possible. More so than their petrol or diesel counterparts, autonomous electric road vehicles should prove easy to customise, easy to personalise and adept at meeting additional human needs and desires. Many new types, many new uses and many new meanings are likely to find their way onto our roads.

Finally, this chapter discussed the concept of mobility and noted the three dimensions which have been proposed for characterising it: spatial, temporal and contextual. It was emphasised that the contextual dimension offered by the friendly neighbourhood robots will almost certainly prove to be different from, and more diverse than, that offered by current automobiles. Automation will make possible interactions which support human needs and desires which transcend simple travel, via additional activities which are today usually within the realm of work, shopping, health or leisure. Due to the advances in automation the friendly neighbourhood robots will inevitably offer a range of interactions with people across different sensory modalities, bridging space and time in ways that were simply not possible with 20th century automobiles.

Having introduced the artefact, i.e. the autonomous road vehicle, the next chapter will introduce the author's preferred approach to designing it: Human Centred Design.

References

ADD HOME Mobility Management And Housing Project 2009, European Union IEE – Intelligent Energy Europe Programme.

Ahangar, M.N., Ahmed, Q.Z., Khan, F.A. and Hafeez, M. 2021, A Survey Of Autonomous Vehicles: enabling communication technologies and challenges, Sensors, Vol. 21, No. 3, p. 706.

Ataya, A., Kim, W., Elsharkawy, A. and Kim, S. 2021, How To Interact With A Fully Autonomous Vehicle: naturalistic ways for drivers to intervene in the vehicle system while performing non-driving related tasks, Sensors, Vol. 21, No. 6, p. 2206.

Carmona, J., Guindel, C., Garcia, F. and de la Escalera, A. 2021, eHMI: review and guidelines for deployment on autonomous vehicles, Sensors, Vol. 21, No. 9, p. 2912.

Centre for Connected and Autonomous Vehicles 2020, Connected And Automated Vehicles: vocabulary v2.0, British Standards Institution, London, UK.

Endsley, M.R., Bolté, B. and Jones, D.G. 2004, Designing For Situation Awareness: an approach to user-centered design, CRC Press, Boca Raton, Florida, USA.

Enge, P., Enge, N. and Zoepf, S. 2021, Electric Vehicle Engineering, McGraw Hill Education, New York, New York, USA.

Eriksson, A. and Stanton, N.A. 2018, Driver Reactions To Automated Vehicles: a practical guide for design and evaluation, CRC Press, Boca Raton, Florida, USA.

Eurostat 2021, Passenger Mobility Statistics, Retrieved from https://ec.europa.eu/eurostat/statistics-explained/index.php?title=Passenger_mobility_statistics#Distance_covered

Groeger, J.A. 2000, Understanding Driving: applying cognitive psychology to a complex everyday task, Psychology Press Ltd, Hove, East Sussex, UK.

Harrow, D., Gheerawo, R., Boyd Davis, S., Phillips, D., Lockton, D., Mausbach, A., Wu, J., Ramster, G., Johnson, S., Meldaikyte, G. and Piliste, P. 2020, Driverless Futures: design for acceptance and adoption in urban environments, Royal College Of Art, London, UK.

Holtzblatt, K., Wendell, J.B. and Wood, S. 2005, Rapid Contextual Design: a how-to guide to key techniques for user-centered design, Morgan Kaufmann Publishers, San Francisco, California, USA.

Jittrapirom, P., Caiati, V., Feneri, A.M., Ebrahimigharehbaghi, S., Alonso González, M.J. and Narayan, J. 2017, Mobility As A Service: a critical review of definitions, assessments of schemes, and key challenges, Urban Planning, Vol. 2, No. 2, pp. 13–25.

Kakihara, M. and Sorensen, C. 2001, Expanding The "Mobility" Concept, SIGGROUP Bulletin, Vol. 22, No. 3, pp. 33–37.

Leon, F. and Gavrilescu, M. 2021, A Review Of Tracking And Trajectory Prediction Methods For Autonomous Driving, Mathematics, Vol. 9, No. 6, p. 660.

Macey, S. and Wardle, G. 2009, H-Point: the fundamentals of car design & packaging, Art Center College of Design, Design Studio Press, Culver City, California, USA.

Mack, A. and Rock, I. 1998, Inattentional Blindness, The MIT Press, Cambridge, Massachusetts, USA.

Marres, N. 2020, Co-Existence Or Displacement: do street trials of intelligent vehicles test society?, The British Journal Of Sociology, Vol. 71, No. 3, pp. 537–555.

Maurer, M., Gerdes, C., Lenz, B. and Winner, H. 2016, Autonomous Driving: technical, legal and social aspects, Springer Nature, Berlin, Germany.

Peacock, B. and Karwowski, W. 1993, Automotive Ergonomics, Taylor & Francis, London, UK.

Pettigrew, S., Fritschi, L. and Norman, R. 2018, The Potential Implications Of Autonomous Vehicles In And Around The Workplace, International Journal Of Environmental Research And Public Health, Vol. 15, No. 9, p. 19–33.

Rosique, F., Navarro, P.J., Fernández, C. and Padilla, A. 2019, A Systematic Review Of Perception System And Simulators For Autonomous Vehicles Research, Sensors, Vol. 19, No. 3, p. 648.

SAE, Taxonomy 2014, Definitions For Terms Related To On-Road Motor Vehicle Automated Driving Systems, J3016, SAE International Standard.

Seba, T. 2018, Rethinking Transportation 2020–2030: disruption, implications & choices, American Public Transportation Association (APTA), Transit CEO Seminar, February 11th, Miami, Florida, USA.

Stanton, N., Revell, K.M. and Langdon, P. eds. 2021, Designing Interaction And Interfaces For Automated Vehicles: user-centred ecological design and testing, CRC Press, Boca Raton, Florida, USA.

Strömberg, H., Pettersson, I., Andersson, J., Rydström, A., Dey, D., Klingegård, M. and Forlizzi, J. 2018, Designing For Social Experiences With And Within Autonomous Vehicles – exploring methodological directions, Design Science, Vol. 4, e13, pp. 1–29.

Thrun, S., Montemerlo, M., Dahlkamp, H., Stavens, D., Aron, A., Diebel, J., Fong, P., Gale, J., Halpenny, M., Hoffmann, G. and Lau, K. 2006, Stanley: the robot that won the DARPA Grand Challenge, Journal of Field Robotics, Vol. 23, No. 9, pp. 661–692.

United Nations 2006, Convention on the Rights of Persons with Disabilities, Retrieved from www.un.org/development/desa/disabilities/convention-on-the-rights-of-persons-with-disabilities/convention-on-the-rights-of-persons-with-disabilities-2.html

Vallance, P. and Norman, J. 2019, A time of unprecedented change in the transport system, The Future Of Mobility, Foresight, UK Government Office For Science, London, UK.

Xing, Y., Lv, C., Cao, D. and Hang, P. 2021, Toward Human–Vehicle Collaboration: review and perspectives on human-centered collaborative automated driving, Transportation Research Part C: Emerging Technologies, Vol. 128, p. 103–199.

Yoon, Y., Chae, H. and Yi, K. 2021, High-Definition Map Based Motion Planning, and Control for Urban Autonomous Driving (No. 2021-01-0098), SAE Technical Paper, Society Of Automotive Engineers, Warrendale, Pennsylvania, USA.

Chapter 3

Human Centred Design

Humanism

Dictionaries suggest that the word "humanism" was first used in the English language in the 19th century. It referred initially to the reengagement with classical antiquity which occurred in Renaissance Italy via Latin works such as those of Cicero, Seneca, Boetius and Saint Augustine (Law 2011). While late medieval education consisted mostly of logic, natural philosophy, medicine, law and theology, the Renaissance humanists developed instead an educational approach based more on history, moral philosophy, grammar, rhetoric and poetry which today often form part of programmes of study in the humanities.

From the early 20th century onwards the word "humanism" has been frequently associated with non-religious and non-aligned secularist movements. As an ideological outlook, humanism usually affirms notions of agency, freedom and progress (Copson and Grayling 2015). It views humans as solely responsible for their own development and emphasises concern for humans within the wider universe.

Humanism usually involves viewing matters from the perspective of people, and placing people at the logical and ethical heart of decision making. The emphasis on individual agency and on the power of social interactions leads to a worldview which Wintermute (2019) has expressed in terms of ten humanist commitments:

- critical thinking;
- ethical development;
- peace and social justice;
- service and participation;
- empathy;
- humility;
- environmentalism;
- global awareness;

DOI: 10.4324/9781003319740-3

- responsibility;
- altruism.

A humanist worldview can significantly influence design since it prioritises human considerations over technological considerations. Humanism and technology-push are not the best of bedfellows. Humanists are naturally drawn to the psychological nature, sociological impact and meaning of the artefact for the people involved. The process and priorities are distinctly different when the human is taken as the datum for the design rather than some other possible point of reference.

And design itself has at times been suggested to be an applied form of humanism. Guellerin (2021) for example suggested that "design is a humanism in the sense that it makes individuals responsible for the world in which they want to live… artistic creation, from which design differentiates itself completely, implies a result which provokes, at least for its author, emotion, pleasure, and values. The designer goes further and introduces the notion of progress for mankind."

Human Centred Design

The word "design" is used by people in different ways. As a noun, dictionaries suggest sketches, drawings, plans, patterns, intentions, purposes or the art of producing them. As a verb, dictionary definitions suggest the representing of an artefact, system or society, or the fixing of its look, function or purpose. The word "design" is therefore used in relation to matters which range from the conceiving of something to the actual plans and processes required to achieve it (Giacomin 2014).

The design approach which is commonly these days referred to as Human Centred Design (HCD) has its roots in fields such as ergonomics (Meister 1999), computer science and artificial intelligence. The echoes of this past can be noted from international standards such as ISO 9241-210 "Ergonomics of human-centred system interaction" which describe HCD as "an approach to systems design and development that aims to make interactive systems more usable by focusing on the use of the system and applying human factors/ergonomics and usability knowledge and techniques".

Such standards tend to address well the needs of the users of tools which have predetermined functions. The difficulty with adopting such a view in the case of road vehicles is that the people involved do not always think of themselves as the "user" of a "tool". Designing for a user usually involves optimising the artefact based on a set of preconceived plans and schema which describe its intended function and use.

31

Such a view tends to lead to designs which are highly efficient towards one or more predetermined usage patterns (Degani 2004) but which are characterised by only limited interactivity, exploration and learning. Such a view is not always well suited to the design of road vehicles.

With highly complex multifunction machines such as road vehicles the use of "personas" and "scenarios" (Mulder and Yaar 2006) has tended to provide greater opportunities for optimising not just the planned functions but also the many other possible interactions (Moggridge 2007) and meanings (Giacomin 2017). Designers use "personas" to make a conscious and often elaborate attempt at clarifying the possible sensations, thoughts and actions of the people which they most wish to satisfy. And designers use "scenarios" to make a likewise determined attempt at clarifying the locations, intentions and interactions which the artefact must be fully compatible with. Choosing between design options is usually easier and better documented when based on evidence of the needs and preferences of the people for whom the design is intended.

Personas and scenarios are in fact typical tools of Human Centred Design as it is currently commonly intended. HCD is based on the use of techniques which communicate, interact, empathise with and stimulate the people involved, obtaining an understanding of their needs, desires and experiences which often transcends that which the people themselves realised beforehand. Practised in its most basic form, human centred design leads to products, systems and services which are physically, perceptually, cognitively and emotionally intuitive (Giacomin 2014). Practised in more advanced form, it becomes a platform for co-design (Sanders and Stappers 2008) and a business model (Osterwalder and Pigneur 2010) based on partnership with the customers or constituency members. It is a form of structured empathy.

Human Centred Design can be represented visually by means of the pyramid which is illustrated in Figure 3.1. It involves the asking of questions which span from the physical nature of people's interaction with the intended product, system or service to the metaphysical. The physical, sensory and cognitive considerations fall mostly within the lower part of the pyramid while the considerations which involve symbolism, value and meaning are mostly in the top part of the pyramid. HCD involves the repeated asking of who, what, where, when and why questions through ethnography, co-design and customer testing. And at each point in the process the answers which the people are providing are used to influence the choices and shape the evolution of the product, system or service. In Human Centred Design the decision making gravitates around people's ideas and opinions more than around the material details of

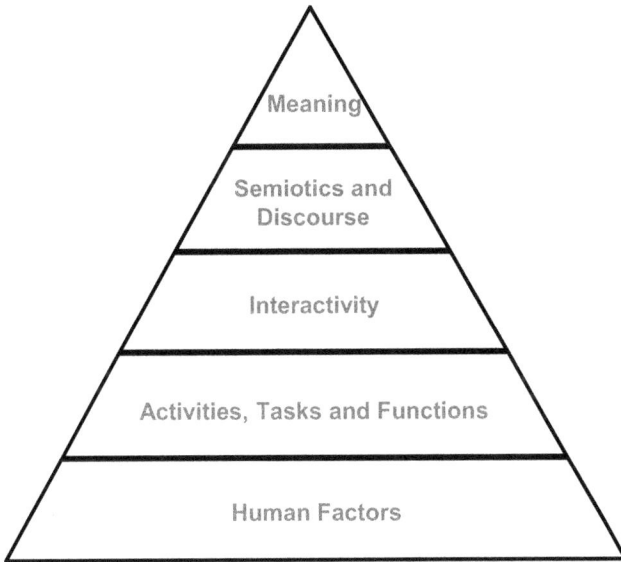

Figure 3.1 Human Centred Design pyramid.

the artefact. HCD focusses on what people prefer more than it does on what the manufacturer or service provider prefers to offer.

Designs which meet their safety, comfort and quality targets but which also address questions and curiosities which are near the top of the HCD pyramid would be expected to embed themselves deeply within people's lives. In particular, the introduction of a new meaning (Giacomin 2017) into a person's life would be expected to offer ample opportunities for commercial success and for brand development, as historic examples from the Piaggio Vespa (Boni and Cordaro 2021) to Ferrari cars (Kapferer and Valette-Florence 2018) seem to suggest.

Human Centred Design is a process within which the characteristics of the artefact are important not in themselves, but as affordances, interactions, semantics and meanings for the people involved. What matters most are not the artefact's physical or informatic characteristics but instead the resulting experiences and meanings. Design changes are judged to be helpful or unhelpful based on the changes which they induce in people's experiences and meanings. The Human Centred Designer needs to be an "old soul".

And one increasingly popular way of evaluating changes in the experiences and meanings is by measuring the human emotions involved. As a global human physiological response which is shaped by the interactions with the designed artefact, human emotion can often be indicative of the success or failure of the design. In fact, practitioners of HCD often specifically target emotional engagement (Jordan 2000; Van Gorp 2012; Chapman 2005; Hill 2010) and measure it to provide evidence of success or failure. Quite simply, few people engage with things or situations which are emotionally unsatisfying, thus evidence of the emotional responses to the given design can provide indirect evidence of success or failure.

Given the human focus, the process of Human Centred Design must necessarily start with the customers or constituency members involved. This differs somewhat from traditional technology-push processes which are usually initiated in a setting such as a research lab where some new material, machine, software or capability can be experimented and developed. Where in the past a design concept may have been ideated by some creative individual through a lightbulb moment, then given physical manifestation through technical development to place a prototype in front of the public for judgement, HCD tends instead to work the other way around.

In Human Centred Design the key is to work with people via ethnography, co-design and customer testing from the first moment of the project. The objective is to identify the metaphors and meanings which are compatible with the constraints and specifications. As shown in Figure 3.2 a stereotypical HCD work flow starts with the target audience, then progresses through concept, engineering, prototyping and validation. Rather than a lightbulb moment setting in motion a series of events, HCD gradually co-designs with at least the major stakeholders such that a successful outcome is guaranteed. With HCD, the progressing through modelling, prototyping, fabrication and validation stages is

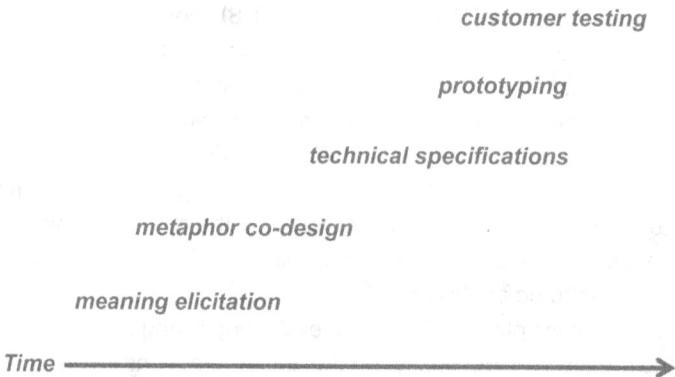

customer testing

prototyping

technical specifications

metaphor co-design

meaning elicitation

Time ───────────────────────────────→

Figure 3.2 Typical sequence of activities in Human Centred Design.

usually performed only after the customers or constituency members are understood well enough to instil confidence in the prospected outcome.

Like any form of design, Human Centred Design is not a science and thus should not be thought of in terms of right or wrong. Perhaps it is also not helpful to think of the methods involved as being more or less accurate. And use of the word "problem", as in "problem solving", should probably also be avoided since a problem is often something which is understood a priori, not necessarily something which emerges on the spot with the people involved. Better, perhaps, to think of Human Centred Design as a form of structured empathy which helps identify "opportunities".

Human Centred Design Tools

The toolbox of Human Centred Design grows continuously, sometimes by borrowing from fields such as psychology or sociology (Berg 2001) and sometimes instead by defining new approaches which emerge from practice. Card decks such as those by IDEO (IDEO 2003) and LUMA (Luma Institute 2012) are fashionable because they are simple tools which make interacting with customers or constituency members easier. Numerous card decks, canvases and frameworks are now available for organising encounters with people for purposes of design, and many have Human Centred Design as the declared philosophical or ideological basis.

One way of thinking about HCD tools is to consider the information-gathering processes involved. The most basic form of HCD tool consists of gathering well-established and well-documented facts about people which are easily accessed via the internet. Many anthropometric, biomechanical, cognitive, emotional, psychophysical, psychological and sociological datasets and models are available today in the research literature. Such items of information, which are often treated as matters of human factors, provide factual statements regarding the abilities and limitations of humans. They tend to be the science behind the design. Such tools define the boundaries within which to operate and usually act more to inform the design process than to drive it.

Another form of HCD tool consists of the gathering of new data, ideas and opinions which are contextually relevant to the project. Some HCD tools consist of verbal or non-verbal methods for interacting with people to detect meanings, needs and desires. Language-based techniques such as ethnographic interviews (Spradley 1979), questionnaires (Hayes 1992) and focus groups (Krueger and Casey 2015) tend to dominate this category historically. Ever more methods are however being deployed to investigate human mental activities which are not accessible in conscious thought or expressible in natural language. Projective techniques (Soley and Smith 2008), participant observation (Spradley 1980), body language analysis (Navarro 2008; Wharton 2009), facial coding analysis (Hill 2010), electroencephalograms (Du Plessis 2011) and other approaches for collecting and analysing non-verbal information are becoming increasingly popular with designers.

Finally, a growing number of HCD tools help detect cultural strategies (Holt and Cameron 2010) and help simulate possible futures. Rather than focus on facts about people and societies, or gather current data, ideas and opinions, these tools help search for patterns and help to hypothesise and experiment new possibilities. The information gathering involves one or more possible futures. From the currently popular approach of co-design (Von Hippel 2005 and Von Hippel 2007) to the more speculative techniques such as real fictions and para-functional prototypes (Dunne 2008), approaches are being deployed to immerse people in possible futures and to socially explore the envisaged world for purposes of design. This set of HCD tools makes possible and documents the gathering of information which is not about the present, but instead about the possible future.

Many well-known methods (Gkatzidou et al. 2021) can be used as HCD tools to get to know people, learn about their needs and desires, and discover patterns of interaction with specific products, systems or services. Any list of methods which can prove useful when performing Human Centred Design runs into the hundreds. The choice of which to deploy is usually based on the characteristics of the target customers or constituency members, the locations, intentions and interactions which characterise the situation, and any specific features, metaphors or meanings which the designers must achieve. The design context is what makes any given method appropriate or inappropriate for use, not its intrinsic physiological, psychological, sociological or mathematical properties. And the cost in terms of time and money required to deploy the method is usually an important practical consideration.

Given the number and variety of potentially useful methods it is not the objective here to discuss all those which might help in learning more about people and about their interactions for purposes of design. Such an exposition would necessarily need to be lengthy if it were to do some degree of justice to the topic. And it would meander beyond the boundaries of the current book. The methods themselves, and ways of choosing the most opportune, are described elsewhere (see for example Gkatzidou et al. 2021).

Only two illustrative examples are presented here due to their ubiquity: "personas" and "scenarios". Each method is a form of summary. Each method integrates disparate known items of information about the person (personas) or the situation (scenarios). Both are non-scientific extensions of the known, sometimes scientific, facts. Both attempt to prioritise those psychological and emotional considerations which are most likely to influence the human experience. Both are based on underlying assumptions and may thus risk missing some motivations and interactions. And both represent essential tools-of-the-trade of Human Centred Design.

It is frequently said that practical design personas were first described in the 1980s by Alan Cooper (2004). Personas are imaginary characters who are, however, based on real facts about real people. Personas consolidate archetypal descriptions of behaviour patterns into representative profiles, and answer the question: "Who are we designing for?"

Three types of information are found in most design personas: demographic, sociological and behavioural. For example, when designing an autonomous road vehicle the demographic information might include age and gender, the sociological information might include job title and education level and the behavioural information might include preferred forms of mobility or attitudes about autonomous road vehicles. The data which is available a priori is usually supplemented by information collected via ethnographic activity. At the end of the data gathering the facts and insights are usually expressed as descriptions that include skills, goals, attitudes, environments and patterns of behaviour (see example in Figure 3.3).

Name: Iona
Age: 72.
Gender: female.
Location: New York City.
Occupation: community services administrator.
Family: widow, two sons and large extended family.
Jungian Architype: explorer.
Hobbies: classical music, jazz music and sewing.
Travel: short commutes to work and long weekend drives in countryside.
Likes: Enjoys taking public transport for people watching, but returns home before dark due to safety concerns. Enjoys large American automobiles and enjoys on-board cellphone integration.
Dislikes: red cars, saying they remind her of fire engines. Does not enjoy the local radio stations and struggles getting groceries in and out of her car.

Figure 3.3 Example of an automotive persona.

Source: 123RF.com

Mulder and Yaar (2006) have suggested that personas are tools which help to build empathy but that to be useful they need to be credible by ensuring that:

- each persona represents real users;
- the personas' attributes and descriptions are accurate and complete;
- the set of personas covers the full range of users.

Despite the near certainty of not managing to capture all the possible behaviours, values and meanings in practice, the personas nevertheless provide a useful tool for anticipating the major expectations and key interactions. And the usefulness increases with increasing ethnographic activity, since more and more insights are noted and accommodated within the set of design personas being developed for the project. Personas provide "virtual team members" whose needs and desires help to evaluate the design options.

A scenario is instead a story describing a sequence of events (Kahn and Wiener 1967), an interaction between people and an artefact (Carroll 2000), or a possible future (Ogilvy and Schwartz 1996). Scenarios consist of a setting or stage with one or more actors who have knowledge, motivations and capabilities. Scenarios describe the actions and activities of each persona in context, including the persona's goals, plans and typical reactions.

Scenarios are usually synthesised from the results of ethnographic activities which investigated the intentional, spatial and temporal characteristics of each environment. They usually include some description of the story background, the motivations, the tasks and the wider context in which the interaction takes place. They can be used to fine-tune the design and to raise questions about the design assumptions. They are tools which provide constraints and which often suggest obvious evaluative criteria.

Scenarios can differ in nature due to differences in what the designers wish to focus on. Functional concerns usually lead to scenarios of the type suggested by Cooper (2004) who claimed to develop mostly "daily-use scenarios" and "necessary-use scenarios". According to Cooper, "daily-use scenarios" capture the actions which the user would be expected to perform most frequently. Such scenarios are suggested to be the most important because they help the designer to consider those aspects of the product, system or service which manifest themselves most frequently. Cooper's second scenario type, the "necessary-use scenario", provides instead a description of actions which must be performable even if only occasionally. For a road vehicle a "breakdown" is an example of a "necessary-use scenario" since the designer hopes that it never happens, but nevertheless must ensure that the needed actions can be performed if it does.

Many functional scenarios are routinely used in automotive design for setting targets and for choosing test environments in relation to engineering matters such as vehicle handling dynamics, vehicle comfort, energy efficiency and crash safety. Functional scenarios have also been much used in relation to matters of ergonomics such as vehicle ingress/egress, visibility from the driver's seat, reachability and boot (trunk) loading and unloading.

In recent times automotive designers have however also shown a growing interest in scenarios which stimulate the human emotions. The emotional states of people are influenced by many of the interactions with the vehicle cabin, vehicle controls and digital communication technologies. Given the importance of human emotions in matters ranging from road safety to purchasing decisions, affective scenarios (Cha 2019) are now sometimes used to consider specifically those situations where the greatest positive or the greatest negative emotions tend to occur. Figure 3.4 provides an example of an automotive affective scenario which naturally stimulates questions about the vehicle's ability to support meaningful events, about the vehicle's integration with social media and about the vehicle's ability to support, or not support, privacy.

A father bought a new car and took it out with his son for a first drive. They enjoyed taking the seat protector sheets off together. After the father pressed the ignition button, his son became excited, playing with every dial in the centre console and touching the satnav, as these were new to him. The father took a picture of his son playing in the car to capture the moment. They then drove to the local coast and parked at the beach for ice cream. The father thought that his son would soon be a man, and wished that he had more than just a picture to remember the day.

Figure 3.4 Example of an automotive affective scenario.

Source: adapted from Cha 2019

Human Centred Design Triangulation

Humans are complex creatures. There is no avoiding the fact that no amount of observation, interaction, conversation or testing is sufficient to fully understand another person's needs and desires. And designing is further complicated by the fact that something like a road vehicle must be considered useable, acceptable and enjoyable by more than a single individual and in more than a single context. The combinations of people and contexts can easily overwhelm the time and resources available, thus the best that can be achieved in practice is to learn as much as possible, and to then hope that it is enough.

Given the certainly that people can never be fully understood or fully described, a question which arises is "then what exact type of information am I actually collecting when I interact with that person in the manner of that method?". If the choice of target customers or constituency members has been made and carefully documented, and if the choice of the locations, intentions and interactions which characterise the situation has also been made and carefully documented, then for those people and those situations what type of information is the method providing?

Table 3.1 provides one possible answer to such questions. It contains a list which was produced as part of a programme of research (Ghatzidou et al. 2021) to identify Human Centred Design methods which were particularly useful in automotive design settings. Focus groups consisting of automotive design experts labelled HCD methods based on whether the gathered information could be considered to be capturing mostly publicly declared positions and publicly performed actions, or, instead, mostly private thoughts, values and beliefs.

As the table suggests, not all of the identified HCD methods were considered to gather both the externally acted declarations and behaviours of the participants (the visible) and their internal thoughts, values and beliefs (the invisible). Some HCD methods seem specifically designed to capture only externally visible actions, while others seem specifically intended to probe the human subconscious to reveal values and beliefs. A few methods appear to gather a mixed bag of both types of information, with it being difficult in practice to estimate which type of information is most prevalent. Psychology and sociology are not mathematics.

If distinctly different types of information can be gathered, and if some methods naturally collect mixed assortments of information types, then what is to be done? If no single Human Centred Design method can guarantee an exhaustive data collection which achieves all

TABLE 3.1 Human Centred Design methods which are particularly useful in automotive design (◉ indicates collecting mostly externally acted declarations and behaviours while ⊘ indicates collecting mostly internal thoughts, values and beliefs).

Method	Ext.	Int.	Method	Ext.	Int.
AEIOU	◉	⊘	Harris Profile	◉	
Affinity Diagram	◉	⊘	Heuristic Evaluation	◉	
Body Storming	◉	⊘	How Might We?	◉	
Brainstorming	◉		Interview	◉	⊘
Buy A Feature	◉	⊘	Laddering		⊘
Card Sorting	◉	⊘	Love/Break Up Letter		⊘
Co-Design	◉	⊘	Persona		⊘
Cognitive Walkthrough	◉		Picture Cards	◉	⊘
Cognitive Map	◉		Repertory Grid Technique		⊘
Competitive Analysis	◉		Role Playing	◉	⊘
Contextual Inquiry	◉	⊘	Scenario	◉	
Crazy 8s	◉		Scenario Mapping	◉	
Crowdsourcing	◉		Stakeholder Analysis	◉	
Cultural Probes		⊘	Storytelling	◉	⊘
Customer Journey	◉		Storyboarding	◉	⊘
Delphi Survey	◉		Survey	◉	
Design Fiction	◉	⊘	Think-Aloud	◉	
Desirability Testing	◉		Touchstone Tour	◉	
Diary Study	◉	⊘	Tomorrow's Headlines	◉	⊘
Empathy Map		⊘	Wizard Of Oz	◉	
Experience Prototyping		⊘	Word Concept Association		⊘
Extreme Users	◉	⊘	Zaltman Metaphor Elicitation		⊘
Fly-On-The-Wall	◉	⊘	5 Whys		⊘
Focus Group	◉				

useful items of information, then what? How does the designer proceed with a degree of confidence?

One approach for reducing some doubts and uncertainties is triangulation. The term is inspired by the perfect number "three" and provides a way forward if two information sets provide opposing indications. With triangulation the designer chooses three HCD methods which are considered to differ in objective, underlying logic and practical detail. The reasoning is that any given HCD method is characterised by specific constraints and biases, thus it seems prudent to gather data using more than a single method. Assembling a set of results which were obtained in a variety of ways helps to reduce the missing information and to note any common themes which appear in more than one of the datasets.

A specific form of triangulation advocated by the author (see Ghatzidou et al. 2021) consists of selecting one HCD method from each of three basic categories: known facts about people, what the person says and what the person does (see Figure 3.5).

With any design activity there will always be some facts about the physical, perceptual, cognitive and emotional constraints which are known a priori from literature. Some facts, often of a scientific nature, are nearly always known about the human and about the context prior to initiating primary research. Such facts form the first leg of the recommended triangulation. Such facts are usually gathered by designers as part of what is sometimes referred to as "desk research" or "secondary research". Reality requires that the design work remain within the envelope of the possible.

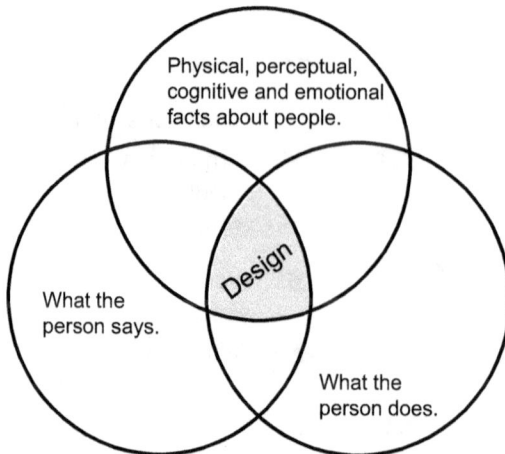

Figure 3.5 The logic of HCD triangulation.

The second leg of the recommended triangulation involves methods for collecting information about what a person or a group "says". An HCD method is selected, with time and resource constraints in mind, which asks questions and gathers the subsequent verbal responses. Questionnaires, interviews, empathy maps, focus groups and several other HCD methods (Ghatzidou et al. 2021) are well suited to linguistic and cognitive inquiry about a given scenario or a given design opportunity.

Consideration of the familiar phrase "do what I say and not what I do" suggests however that it is probably unwise to rely too much on statements. The cognitive, rational, reasoning which is expressed via natural language is known (Blackmore and Troscianko 2018; Loftus 1996) to be more of a story than a factually accurate description of the activities and thoughts which had occurred. It would thus seem wise to reduce the risk of misunderstandings by adopting a third leg which involves an HCD method that captures mainly the actions which occur when people are exposed to the design scenario or to the design opportunity.

The third leg of the author's recommended triangulation collects information about what a participant or a group of participants does in the prospected situation. Time-and-motion analysis, cognitive walk-throughs, fly-on-the-wall observation, body motion tracking, button click tracking, eye tracking and other techniques and technologies can be used to record what occurs (see Gkatzidou et al. 2021 for examples). While not always fully accurate or fully repeatable, such information nevertheless tends to be more physically grounded and less open to interpretation than the ideas or opinions expressed by participants. While ideas and opinions can often prove somewhat uninhibited, interactions and behaviours tend to implicitly respect the real-world constraints which are present in the situation.

But of course, with any categorisation or taxonomy, there are always real-world examples which do not fit conveniently into the established categories or which fit into more than one simultaneously. And the recommended triangulation procedure is no exception. Few HCD methods produce pure "a priori facts" or pure "what the person says" or pure "what the person does". As with any choice, that of the methods to use must inevitably be based on multiple considerations including the available time, the available resources, the intended use of the information and the degree of priority set by the sponsors or designers. As always, a degree of human judgement is required.

Facial expression analysis (Akamatsu 2019; Ekman and Friesen 2003; Meiselman 2016) is a good example of the method selection complexities since the measurements are of externally visible facial expressions and externally visible expressive movements, while what is usually inferred from the data are the internal emotional states of the individual. In a way, the technology directly measures "what the person does" but aims to reveal something closer to "what the person says".

Co-design is another interesting case since there is no certain manner for knowing whether the contributions of a given individual during a co-design process are more the result of internal convictions or, instead, the external social pressures of the setting. Co-design offers people multiple opportunities for stating or even acting out needs and desires, thus the approach is difficult to fit into such a simple taxonomy as that of "says" or "does".

Human Centred Design Of Autonomous Road Vehicles

Automotive design has evolved continuously from the time of the first horseless carriages of the 19th century. While the changes in technology, priorities and working methods do not follow a simple pattern, attempts have nevertheless been made to categorise the periods for purposes of classification and understanding. One interesting proposal was put forward by Jaafarnia and Bass (2011) who identified seven automotive eras:

- Invention Era (1885–1896): designers focussed on achieving a horseless carriage capable of running entirely on its own power.
- Innovation Era (1896–1908): designers noted that the form of a horse-drawn carriage was not ideal and began attempting different arrangements based on the mechanical and commercial possibilities.
- Manufacturing Era (1908–1915): designers focussed on simplification and standardisation in order to facilitate manufacture, following the lead set by Ford Motor Company.
- Capsule Era (1920–1930): designers started using the space within the cabin to modify the value and the meaning of the vehicle. Automotive design grew into a specialised discipline.
- Classic Era (1930–1940): designers moved away from rectangular body shapes in order to improve the aerodynamics, following the lead set by motor-racing, trains and aeroplanes.
- Integration Era (1949–1968): designers focussed on joining the structural parts of the car into a unibody and on increasing safety and comfort. Marketing professionals often emphasised performance and sportiness.
- Modern Era (1968–2008): in response to market forces designers improved functionality and developed new formats such as

hatchbacks, minivans and sport utility vehicles. The design process itself increasingly incorporated computer-based technologies such as CAD, FEA and virtual reality.

In recent years automotive manufacturers and many other organisations have tended to prioritise the experiences (Chapman 2005; Schifferstein and Hekkert 2007; Shaw et al. 2010) which their products, systems or services offer. Safety, comfort and quality have become minimum requirements, while enjoyable experiences have come to be expected. The design emphasis has thus shifted even further in the direction of the people, and it might be said that we have been living through an Experience Era of automotive design.

And in recent years the Human Centred Design approach has been increasingly deployed as part of the effort to achieve those optimal experiences. HCD methods have been deployed to explore the interactions which occur and the meanings (Verganti 2009; Giacomin 2017) and ethical considerations involved (Brown 2005; Arnold 2009). Given the vast number of contextual factors and road vehicle characteristics which can affect the experiences, the HCD approach has been helping to reduce the design complexity down to a small enough number of issues and parameters to deal with. The human lens has been helping to reduce the infinite to the finite.

But why talk at length about Human Centred Design when discussing autonomous road vehicles? Is there some similarity of characteristics or some natural affinity between the manner of the designing and the artefact which is being designed? One reason why the two topics seem appropriate bedfellows is the current state of autonomous road vehicle development, and its current maturity as a paradigm. At the moment the autonomous road vehicle is a new opportunity which is in search of a purpose and of public acceptance. And Human Centred Design is well suited to just those tasks.

Automotive design has traditionally been heavily constrained by the physics involved (see for example Macey and Wardle 2009 or Meadows 2018). The size and complexity of the internal combustion engines and drivetrains, the control actions needed from the driver, the ingress and egress requirements, and the subtle but important needs of aerodynamics and fuel efficiency have all led to highly constrained design briefs. And many characteristics of 20th century road vehicles were the result of the absolute ergonomic requirements of the driver's position, such as limiting the control inputs to what a driver could safely monitor and operate in real time. Human cognitive workload (Sweller 1988; Wickens

2008) in particular proved to be critical towards guaranteeing driving safety and comfort. The results of these physical and ergonomic constraints can be seen in the many physical and functional similarities of the automobiles of today.

As noted in the previous chapter, however, autonomous road vehicles may prove to be different. Autonomous road vehicles, particularly the autonomous electric ones, seem to be relaxing many traditional constraints. With electric propulsion the reduction in the number of moving parts, their simplification, and their possible redistribution around the vehicle (Seba 2018) all provide new freedoms. And the lack of a dedicated driver's position and of some or all of the human driving controls is a substantial simplification. Finally, the transfer of many other human inputs and operations to the vehicle's automation provides immense benefits in terms of reducing cognitive workload (Sweller 1988; Wickens 2008). Even if only a small fraction of the predicted simplifications and cost reductions (Seba 2018) were realised in practice, the increased automation alone would prove sufficient to achieve much greater tailoring of the road vehicle to fit the needs of people.

Given these new opportunities it is likely that autonomous road vehicles will soon reach our streets which will diverge in important ways from the 20th century metaphors of "automobile", "van" or "lorry". Many truisms about the physical layout of road vehicles are now open for discussion and many marketing segmentations and business models are in urgent need of review. Something new is afoot.

It is thus the autonomous road vehicle's current lack of maturity as a paradigm which naturally invites the use of Human Centred Design. The autonomous road vehicle is something new which needs defining and which requires societal acceptance. And HCD methods such as those listed in Table 3.1 provide a wealth of opportunities for poking and prodding people during the design process, ensuring that issues and opinions emerge along the way rather than after the fact. And where completely new and potentially confusing interactions are being planned, sophisticated co-design (DeLaFlor 2020) approaches in particular can help to probe the possibilities for rendering those interactions simple, helpful and meaningful.

The autonomous road vehicle is currently an opportunity in search of a purpose and of public acceptance. The friendly neighbourhood robots will enter service as new and largely unfamiliar entities. They will perform some traditional transport functions more efficiently than current road vehicles and will probably also offer some new services via their on-board automation. Given the new opportunities provided by the combined innovations of electrification and automation, the design of the friendly neighbourhood robots is likely to be constrained more by the psychological and sociological impacts on people than by the traditional physical and ergonomic constraints. Autonomous road vehicles and Human Centred Design would thus seem to be convivial fellow travellers which are helping us to arrive at our ultimate destination, the Autonomous Era.

Conclusions

Human Centred Design is based on the use of techniques which communicate, interact, empathise and stimulate the people involved, obtaining an understanding of their needs, desires and experiences. It is a form of structured empathy. This chapter has provided a short introduction to humanism, to Human Centred Design, to the tools of HCD, to HCD triangulation and to the Human Centred Design of autonomous road vehicles.

The HCD pyramid was introduced to visualise the issues which are investigated, discussed and prioritised. Physical, sensory and cognitive considerations fall mostly within the lower part of the pyramid while the considerations which involve symbolism, value and meaning are mostly in the top part of the pyramid. Designs which meet their safety, comfort and quality targets but which also address questions and curiosities which are nearer to the top of the HCD pyramid would be expected to embed themselves deeply within people's lives.

It was noted that research studies have identified several hundred methods from design, psychology, sociology and other disciplines which can prove helpful when performing Human Centred Design. It was further noted that some HCD methods seem specifically useful for capturing externally visible actions, while others seem specifically intended to probe the human subconscious to reveal values and beliefs. A few HCD methods appear to gather a mixed bag of both types of information, with it being difficult in practice to estimate which type of information is most prevalent.

The approach called triangulation was introduced to indicate one possible way of dealing with the inevitable uncertainties and gaps in the data which can be gathered from any one form of interaction with people. Each HCD method has strengths and weaknesses, and tends to gather information which is more of one type than of another. Triangulation provides a way of widening the insights and reducing the uncertainties.

Of the many HCD tools only the highly popular methods of personas and scenarios were discussed in detail. Designers use "personas" to clarify the possible sensations, thoughts and actions of the people which they most wish to satisfy. And designers use "scenarios" to clarify the locations, intentions and interactions which the artefact must be fully compatible with. Choosing between design options is usually easier and better documented when based on evidence of the needs and preferences of the people for whom the design is intended. Summaries of the characteristics and attitudes of the people involved (personas) and of the contexts in which the interactions take place (scenarios) are essential for any design process which intends to place people at its heart.

Finally, it was noted that the autonomous road vehicle's current lack of maturity as a paradigm naturally invites the use of Human Centred Design. The autonomous road vehicle is something new which needs defining and which requires societal acceptance. It is currently an opportunity in search of a purpose and of public acceptance. And Human Centred Design offers a wealth of opportunities for ensuring that key issues and opinions emerge during the design process rather than after the fact. It is the author's conviction that, with time, HCD will in fact achieve friendly neighbourhood robots which meet public expectations.

Having introduced both the artefact and the recommended approach for designing it, the next chapter discusses one of the biggest design decisions which affects users, that of leveraging or not leveraging the human anthropomorphising tendency.

References

Akamatsu, M. 2019, Handbook Of Automotive Human Factors, CRC Press, Boca Raton, Florida, USA.

Arnold, C. 2009, Ethical Marketing And The New Consumer, John Wiley & Sons, Chichester, West Sussex, UK.

Berg, B.L. 2001, Qualitative Research Methods For The Social Sciences, Allyn & Bacon Publishers, Boston, Massachusetts, USA.

Blackmore, S. and Troscianko, E.T. 2018, Consciousness: an introduction, Routledge, Abingdon, Oxfordshire, UK.

Boni, V. and Cordaro, S. 2021, Vespa: style and passion, Motorbooks, Quarto Publishing Group, Beverly, Massachusetts, USA.

Brown, M.T. 2005, Corporate Integrity: rethinking organizational ethics and leadership, Cambridge University Press, Cambridge, UK.

Carroll, J. 2000, Five Reasons For Scenario-Based Design, Interacting With Computers, Vol. 13, No. 1, pp. 43–60.

Cha, K. 2019, Affective Scenarios In Automotive Design: a human-centred approach towards understanding of emotional experience, Doctoral Dissertation, Brunel University, UK.

Chapman, J. 2005, Emotionally Durable Design: objects, experiences and empathy, Earthscan Publishers, London, UK.

Cooper, A. 2004, The Inmates Are Running The Asylum: why high-tech products drive us crazy and how to restore the sanity, Vol. 2, SAMS Publishing, Indianapolis, Indiana, USA.

Copson, A. and Grayling, A.C. eds. 2015, The Wiley Blackwell Handbook of Humanism, First Edition, John Wiley & Sons, Ltd., Chichester, UK.

Degani, A. 2004, Taming HAL: designing interfaces beyond 2001, St. Martin's Press, Palgrave Macmillan, New York, New York, USA.

De La Flor Aceituno, D. 2020, Co-Designing With Drivers: a human-centred approach towards the implementation of real-time contextual interviewing, Doctoral Dissertation, Brunel University, UK.

Dunne, A. 2008, Hertzian Tales: electronic products, aesthetic experience, and critical design, MIT Press, Cambridge, Massachusetts, USA.

Du Plessis, E. 2011, The Branded Mind: what neuroscience really tells us about the puzzle of the brain and the brand, Kogan Page Publishers, London, UK.

Ekman, P. and Friesen, W.V. 2003, Unmasking The Face: a guide to recognizing emotions from facial clues, Malor Books, Cambridge, Massachusetts, USA.

Giacomin, J. 2014, What Is Human Centred Design?, The Design Journal, Vol. 17, No. 4, pp. 606–623.

Giacomin, J. 2017, What Is Design For Meaning?, Journal of Design, Business & Society, Vol. 3, No. 2, pp. 167–190.

Gkatzidou, V., Giacomin, J. and Skrypchuk, L. 2021, Automotive Human Centred Design Methods, Walter de Gruyter GmbH, Berlin, Germany.

Guellerin, C. 2021, Design, Ethics, And Humanism, Design Management Review, Vol. 32, No. 2, pp. 44–49.

Hayes, B.E. 1992, Measuring Customer Satisfaction: development and use of questionnaires, ASQC Quality Press, Milwaukee, Wisconsin, USA.

Hill, D. 2010, Emotionomics: leveraging emotions for business success, 2nd Ed, Kogan Page Ltd, London, UK.

Holt, D. and Cameron, D. 2010, Cultural Strategy: using innovative ideologies to build breakthrough brands, Oxford University Press, Oxford, UK.

IDEO 2003, IDEO Method Cards: 51 ways to inspire design, W. Stout Architectural Books, San Francisco, California, USA.

Jaafarnia, M. and Bass, A. 2011, Tracing The Evolution Of Automobile Design: factors influencing the development of aesthetics in automobiles from 1885 to the present, In Proceedings of the International Conference On Innovative Methods In Product Design (IMProVe) 2011, Venice, Italy, June 15th–17th, pp. 8–12.

Jordan, P.W. 2000, Designing Pleasurable Products: an introduction to the new human factors, Taylor & Francis, London, UK.

Kahn, H. and Wiener, A.J. 1967, The Next Thirty-Three Years: a framework for speculation, Daedalus, Vol. 96, No. 3, pp. 705–732.

Kapferer, J.N.M. and Valette-Florence, P. 2018, The Impact Of Increased Brand Penetration On Luxury Desirability: a dual effect, Journal Of Brand Management, Vol. 25, No. 5, pp. 424–435.

Krueger, R. and Casey, M.A. 2015, Focus Groups: a practical guide for applied research, Sage Publications Inc., Thousand Oaks, California, USA.

Law, S. 2011, Humanism: a very short introduction, Oxford University Press, Oxford, UK.

Loftus, E.F. 1996, Eyewitness Testimony, Harvard University Press, Cambridge, Massachusetts, USA.

LUMA Institute, 2012, Innovating for People: planning cards, LUMA Institute, Pittsburgh, Pennsylvania, USA.

Macey, S. and Wardle, G. 2009, H-Point: the fundamentals of car design & packaging, Art Center College of Design, Design Studio Press, Culver City, California, USA.

Meadows, J. 2018, Vehicle Design: aesthetic principles in transportation design, Routledge, New York, New York, USA.

Meiselman, H.L. ed. 2016, Emotion Measurement, Woodhead Publishing, Duxford, UK.

Meister, D. 1999, The History Of Human Factors And Ergonomics, Lawrence Erlbaum Associates, Mahwah, New Jersey, USA.

Moggridge, B. 2007, Designing Interactions, MIT Press, Cambridge, Massachusetts, USA.

Mulder, S. and Yaar, Z. 2006, The User Is Always Right: a practical guide to creating and using personas for the web, New Riders Publishers, Berkeley, California.

Navarro, J. 2008, What every BODY is saying: an ex-FBI agent's guide to speed-reading people, HarperCollins, New York, New York, USA.

Ogilvy, J. and Schwartz, P. 1996, Plotting Your Scenarios. In Ogilvy, J. (Ed.), Facing The Fold: essays on scenario planning (pp. 11–34), Triarchy Press, Axminster, Devon, UK.

Osterwalder, A. and Pigneur, Y. 2010, Business Model Generation: a handbook for visionaries, game changers, and challengers, John Wiley & Sons, Hoboken, New Jersey, USA.

Sanders, E.B.N. and Stappers, P.J. 2008, Co-creation And The New Landscapes Of Design, CoDesign, Vol. 4, No. 1, pp. 5–18.

Schifferstein, H.N.J. and Hekkert, P. 2007, Product Experience, Elsevier, Amsterdam, The Netherlands.

Seba, T. 2018, Rethinking Transportation 2020–2030: disruption, implications & choices, American Public Transportation Association (APTA), Transit CEO Seminar, February 11th, Miami, Florida, USA.

Shaw, C., Dibeehi, Q. and Walden, S. 2010, Customer Experience: future trends & insights, Palgrave Macmillan, Basingstoke, Hampshire, UK.

Soley, L. and Smith, A.L. 2008, Projective Techniques For Social Science And Business Research, Southshore Press, Milwaukee, Wisconsin, USA.

Spradley, J.P. 1979, The Ethnographic Interview, Holt, Rinehart and Winston, New York, New York, USA.

Spradley, J.P. 1980, Participant Observation, Holt, Rinehart and Winston, New York, New York, USA.

Sweller, J. 1988, Cognitive Load During Problem Solving: effects on learning, Cognitive Science, Vol. 12, No. 2, pp. 257–285.

Van Gorp, T. 2012, Design For Emotion, Morgan Kaufmann, Waltham, Massachusetts, USA.

Verganti, R. 2009, Design-Driven Innovation: changing the rules of competition by radically innovating what things mean, Harvard Business Press, Boston, Massachusetts, USA.

Von Hippel, E. 2005, Democratizing Innovation, MIT Press, Cambridge, Massachusetts, USA.

Von Hippel, E. 2007, An Emerging Hotbed Of User-Centered Innovation, Breakthrough Ideas For 2007, Harvard Business Review, Article R0702A, February.

Wharton, T. 2009, Pragmatics and non-verbal communication, Cambridge University Press, Cambridge, UK.

Wickens, C.D. 2008, Multiple Resources And Mental Workload, Human Factors, Vol. 50, No. 3, pp. 449–455.

Wintermute, K. 2019, Living Humanist Values: the ten commitments, Humanist, September–October, American Humanist Association (AHA), Washington D.C., USA.

Chapter 4

Anthropomorphism

Anthropomorphism

Dictionaries suggest that the word anthropomorphism refers to the attribution of human traits, emotions, or intentions to non-human entities. It is a tendency to assign human characteristics to inanimate objects or to animals, with a view to rationalising their actions (Duffy 2003).

Mithen (1998) has suggested that anthropomorphism is a defining characteristic of Homo Sapiens Sapiens and that the trait may have developed early in evolutionary terms. Serpell (in Daston and Mitman 2005) has further suggested that "Anthropomorphism appears to have its roots in the human capacity for so-called reflexive consciousness – that is, the ability to use self-knowledge, knowledge of what it is like to be a person, to understand and anticipate the behaviour of others". It appears to be an evolutionary tool which helps us to understand the environment around us and to reduce the time needed for decision making. It is not difficult to imagine ways in which such an innate capability might have assisted our survival as a species.

Surviving fragments of the writings of Xenophanes from around about 500 BCE describe the Greek gods as anthropomorphisms. He claimed that, struggling to imagine differently, people were taking humans as the models for their gods. More than a millennia later people were still reflecting upon such tendencies when Francis Bacon (1620) wrote: "Science has been enormously messed up by this appeal to final causes, which obviously come from the nature of man rather than from the nature of the world, that is, which project the scientist's own purposes onto the world rather than finding purposes in it."

Perhaps the tendency towards anthropomorphising was most succinctly described by the philosopher Davie Hume (1757) a little more than a century later when he wrote:"There is an universal tendency among mankind to conceive all beings like themselves, and to transfer to every object, those qualities, with which they are familiarly acquainted,

 DOI: 10.4324/9781003319740-4

and of which they are intimately conscious. We find human faces in the moon, armies in the clouds; and by a natural propensity, if not corrected by experience and reflection, ascribe malice or good-will to everything, that hurts or pleases us."

Over the centuries the anthropomorphising tendency has been noted to be innate, instantaneous and requiring effort and logic to counteract. It has also been noted to be immediately available to children, perhaps from birth, and to depend to some extent on the object which is being anthropomorphised and on the person who is doing the anthropomorphising. The complexity of the human tendency was articulated by Waytz et al. (2010) when they wrote:"A moment's reflection, however, makes it clear that some agents are anthropomorphised more than others, some cultures seem more prone to anthropomorphism than others, children are generally more likely to anthropomorphise than adults, and some situations increase the tendency to anthropomorphise compared to others. Anthropomorphism is not an invariant feature of everyday life to be taken for granted but rather a wide-ranging and variable psychological process to be explained."

In recent times neuroscience has identified circuitry which contributes to the anthropomorphising tendency and has noted that the circuitry is present from early embryonic stages of human development. Specialised subcortical and cortical circuits facilitate the detection of stimuli such as faces, voices and the emotions which are typical of living creatures. And, as neuronal circuits rather than synaptic strengths, the claim can be made that humans are physically hardwired to some degree for the detection of such patterns and causalities, assisting survival.

Given the wealth of specialised neuronal circuits and their importance in the growth and development of a person it is not surprising that these circuits should deploy subconsciously in nearly all situations, including those for which they are not fully appropriate. Reacting to the events around us as if they were "caused" or "intentional" is a natural part of the human condition.

This leads to well-documented human behaviours such as interacting with inanimate televisions and computers as if they were people (Reeves and Nass 1996). Or the frequent observations in psychological and sociological studies that people ascribe humanlike characteristics to inanimate objects more frequently when undergoing conditions of loneliness and reduced social interaction (Bartz et al. 2016). In its strongest manifestation it leads to attitudes and behaviours which are best described as "personification", where an object or living creature is treated as a person. A tendency which Turner (1987) has attributed to the fact that "We are people. We know a lot about ourselves. And we often make sense of other things by viewing them as people too".

Anthropomorphism is thus a relational filter whose effects are widespread and significant. The anthropomorphising tendency has been noted to be innate, instantaneous and requiring effort and logic to counteract. It can't be ignored.

Anthropomorphising Animals

The anthropomorphising of animals and of animal behaviours is likely to have begun early in human history. Cave drawings and artefacts from the Upper Palaeolithic have been found which seem to show animals with human heads or humans with animal heads or limbs. While difficult from the distance of today's world to understand the reasons for such zoomorphic or anthropomorphic representations, there is little doubt that they were widespread in ancient cultures such as the Egyptian Kingdoms.

And from classical antiquity there are numerous surviving writings which discuss domesticated pets such as dogs and cats in anthropomorphic terms (Campbell 2014). In storytelling, literature and poetry the use of anthropomorphism was well established by then, as exemplified by Aesop's Fables (Temple 2003) and other well-known works in which animals were assigned human values and human behaviours for purposes of narration and moral education.

In more recent times it has also been suggested that the medium itself of language is an anthropomorphising force. Most natural languages are human centric, value laden and gendered. The medium is not an impartial and independent tool, it is instead used in structured ways for specific purposes. For example, Crist (2010) has suggested that anthropomorphising helps writers to express their ideas because such descriptions are fully obvious and fully innate to humans. Crist went on to suggest that the writings of naturalists such as Charles Darwin, ethologists such as Konrad Lorenz and sociobiologists such as Edward O. Wilson differ significantly in terms of the nature and degree of anthropomorphising due to viewing the animals through different interpretative lenses: intentionality, scientific observation or evolutionary social forces. Language usage thus adds an additional layer of complexity to the instinctual human anthropomorphising tendency.

Anthropomorphised animals are very common in storytelling today. Books, cinema, television and the internet are populated by a breathtaking variety of animals which are easily understood and easily stereotyped due to the leveraging of human characteristics by the storytellers.

Childhood Disney favourites greatly expanded the mass anthropomorphising of animals, but even today's most scientific documentaries make obvious use of anthropomorphising when telling the stories of the lives and destinies of the animals which were observed.

Today's books, cinema, television and the internet are also populated by an assortment of hybrid creatures which exhibit both animal and human features. From cartoon favourites to the mascots deployed by sports teams, hybrids are everywhere. Hybrids are so pervasive that their design has become a topic of study in its own right, with one review by Bliss (2012) suggesting a framework involving the mixing of animal form, human form, psychological intent, cultural knowledge and formal blending process.

Anthropomorphising Machines

With respect to the anthropomorphising of animals, that of machines and of machine behaviours seems to be a more recent phenomenon. While this may simply be the result of a scarcity of historical evidence which perhaps distorts analysis, it could instead be due to the growth in complexity of the machines themselves. Ancient tools of the stone and iron ages were rudimentary, not lending themselves to anthropomorphic interpretations without additional decoration or styling. Records from classical antiquity provide some evidence of hand tools, swords or other manufactured artefacts being assigned names and the occasional narrative, but the phenomenon does not seem to have been extensive. Most frequently, the narrative focus seems to have been placed on the nature and exploits of the owner.

Even the early machines of the industrial revolution were rather limited in their complexity and do not seem to have been the subject of sustained anthropomorphising in the manner of today's ships, automobiles or aeroplanes. Their functions, despite being impressive, were often self-evident. And a machine which is characterised by a high degree of repeatability and predictability does not naturally lend itself to anthropomorphising (Waytz et al. 2010). It may be the case that the anthropomorphising, and societal interest in the anthropomorphising, was not fully stimulated until the arrival of the complex multifunction machines of the late 19th and early 20th centuries.

Since the early 20th century, however, anthropomorphism has been more common than is generally appreciated. In design circles the matter of anthropomorphism is often considered due to its impact on product intuitiveness, pleasure and acceptance. Supporting or instead obstructing the anthropomorphising tendency via the characteristics of the artefact is an important design decision. The choice of a degree of anthropomorphism carries with it psychophysical, psychological, social and cultural baggage.

For example there has been interest in pareidolia, i.e. interest in the tendency to see faces on objects. Wodehouse et al. (2018) reviewed more than 2000 images of products found on the internet and concluded that more than half of them exhibited a degree of pareidolia. And since the early 20th century nearly all road vehicles have seemed to have a face, of sorts. It is not difficult to detect a set of eyes and a mouth at the front and back of nearly every one. And houses and buildings are often highly symmetrical with central doorways for access, which produces a face of sorts. Characteristics which approximate a face occur regularly thus many designers seem to be leveraging the opportunity, intentionally or unintentionally, in their work.

Automotive anthropomorphism has been a frequent topic of investigation. For example, Maeng and Aggarwal (2018) reported the effects of face ratio (ratio of face width to face height) on the sense of dominance which people perceive. The research suggested that automobile faces are perceived in mostly the same manner as human faces, i.e. that a higher face ratio is associated with a greater sense of dominance. And a quick look around at most current sport utility vehicles would seem to suggest a current societal preference for projecting dominance.

In another study, Pazhoohi and Kingstone (2020) concluded that larger vehicles such as lorries are usually perceived as more dominant, angry, hostile and masculine than smaller vehicles like automobiles, and that the perceptions can be accentuated with the increasing age of the human observer. Again, a quick look at today's sport utility vehicles seems to suggest a current customer preference for angry and aggressive looking vehicles.

Benfield et al. (2007) noted that drivers tend to assign human qualities such as a name and gender to their road vehicles, attributing to them a form of personality. The researchers also noted that drivers were not projecting their own personalities onto the road vehicles but that the assigned vehicle personality was a strong indicator of the aggressiveness of the driving style of the vehicle owner.

Miesler (2011) considered instead the effect of the usage context on the way people anthropomorphise automobiles. Participants were asked to rank automobiles from within a specific usage context which was either highly functional in nature or instead highly emotional-hedonic in nature. The results suggested that the participants had preferred anthropomorphic car designs over more neutral designs when expressing their opinions from within the emotional-hedonic usage scenario. The degree of the anthropomorphising was found to be a function of the usage context, and it can be argued that the degree of emotional activation was a factor.

Perhaps the most interesting examples of machine anthropomorphism are those which can be found in the world of robotics. A substantial literature considers the effects of design choices which can support or instead obstruct the anthropomorphising tendency. For example, there have been many investigations about the effects of robot physical appearance.

Studies have repeatedly noted (see for example Hiroi and Ito 2008) that people become more anxious and express greater fear when interacting with large robots, particularly those which are larger than the average human. Humanoid robots such as ASIMO or HBS-1 (see Wu et al. 2016) are thus sized similar to an average nine-year-old boy. Honda selected a height of 120 cm for ASIMO due to it being a reasonable compromise between the need to reach door knobs and light switches on the one hand, and the need to not stimulate anxiety or fear in people on the other.

Studies have also repeatedly noted (see for example Bernotat et al. 2017) that more contoured body shapes stimulate female connotations and associations while more angled and squared body shapes stimulate instead male connotations and associations. In addition, the perceived body shapes often stimulate cognitive stereotypes, leading for example to judging a female-shaped robot to be more appropriate for jobs which are stereotypically performed by women and male-shaped robots for jobs which have traditionally been performed by men. And a study by Nomura (2017) further reported that even simple gendering of the robot's name or voice was sufficient to stimulate stereotypes and to change people's reactions to the robot.

A much investigated area has been the design of human-like faces for facilitating interactions (see for example Hara and Kobayashi 1995). The underlying assumption has been that face-to-face conversations between humans are obvious sources of inspiration for human–machine interactions. The key point is that human face-to-face conversations involve a multiplicity of channels including sounds, mouth movements, eye movements, head movements, body postures and other perceptual, cognitive and emotional indicators of intention and meaning. The synchronised postures and movements provide a degree of redundancy and a variety of information sources which help to decode the communication and to support the interpretation of intention and meaning.

Duffy (2003) reviewed the opportunities for supporting social and emotional interactions between robots and humans by means of anthropomorphising. The review produced a set of design guidelines which included:

- avoid the uncanny valley (ensure that the robot is obviously per-ceived as a robot);
- balance form and function (ensure strong correlations between the robot's form and its functions);
- use natural motion (ensure motions are consistent with familiar examples from biological systems such as humans);
- adopt social communication conventions (ensure that the robot coordinates its reactions and interactions in the manner of familiar biological systems such as humans);
- communicate emotions (ensure emotional expressions are used which are similar to those of familiar biological systems such as humans so as to facilitate social interactions);
- facilitate the development of the robot's own identity (ensure inte-gration into the intended social space and the establishment of a unique identity for the robot);
- establish autonomy (ensure that the robot's autonomy is consistent with its capabilities and social roles).

Among the more interesting considerations in relation to machine anthro-pomorphism is the concept of the "uncanny valley" (see Figure 4.1). Mori (1982) suggested that as the appearance of a robot is made more human the people's emotional responses become more positive and

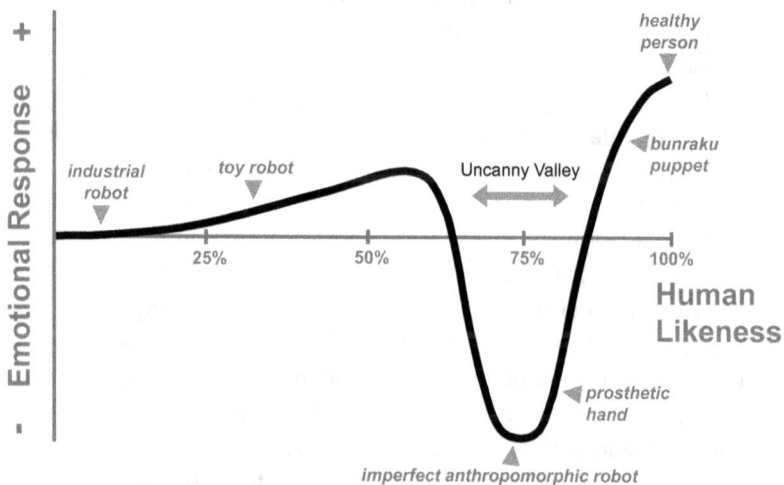

Figure 4.1 The concept of "uncanny valley" in robotics.

empathetic, until a point is reached beyond which they react negatively to the robot due to its imperfect mimicking of humans. Further realism then eventually reaches a point at which the robot becomes indistinguishable from an actual human, restoring positive responses and empathy. The "uncanny valley" of lowered human acceptance has been frequently confirmed in research settings and it provides a useful reference point for separating those design interventions which aim to make a robot more easily interpretable from those which intend instead to mimic humans or other living creatures.

And a particularly noteworthy example of designed anthropomorphism is that of the Tamagotchi. The device was a small egg-shaped handheld digital pet which was created in Japan and sold worldwide from 1996 onwards. It usually had an interface consisting of three buttons which could be used to care for the device via feeding, playing and disciplining. Insufficient care from the owner could cause the digital pet to die and it could also eventually die of old age. It underwent the stages of baby, child, teenager and adult during its lifetime.

The explosion in Tamagotchi sales led to a global phenomenon involving anything from school bans of the device on the one hand to physical cemeteries for dealing with dead devices on the other. By requiring feeding, playing and disciplining inputs from its owner, and by providing signs of emotions and well-being in response, the device invited intense social interactions despite its highly simplified form. The term "Tamagotchi Effect" has in fact come to describe how people develop ritualistic interactions with, and emotional attachments to, inanimate objects in response to anthropomorphic prompting.

Allison (2006) has suggested that the Tamagotchi became "a person's constant companion almost more than anything outside the body itself, offering distraction from the intricacies and intimacies of daily existence". And Lawton (2017) claimed that "The constant presence of the device allowed for it to not only infiltrate the previously fixed social spaces of the user, but also to create new ones in the absence of social interaction".

The history of the Tamagotchi provides a particularly insightful illustration of the power of anthropomorphism and of the opportunities for stimulating the human anthropomorphising tendency via interactions and behaviours. The emotional attachments found in studies of the Tamagotchi were often as strong as any formed with living creatures or with other humans. A fact which most designers should take care to consider.

Scrutiny of the world around us confirms that many complex multifunction machines have anthropomorphic elements, or are holistically anthropomorphised. Faces appear everywhere and a multitude of machine shapes and motions are matched, intentionally or unintentionally, to relevant human shapes and motions so as to facilitate interactions and render them more natural. Given the richness of human neural circuitry for recognising faces, movements and emotions it is perhaps not surprising that so much of the human-made world reflects these subconscious anthropomorphising abilities.

Studies of machine anthropomorphising have suggested that the phenomenon is common, that it can be stimulated in different ways, that it can be person dependent, that it can be context dependent and that it can depend on emotional state. As a natural aspect of human life the anthropomorphising tendency seems unavoidable when interacting with complex multifunction machines. Public acceptance of autonomous road vehicles will thus probably depend to some extent on the degree of anthropomorphism which is stimulated and on the matching of the anthropomorphising characteristics to the intended role of the machine. Autonomous road vehicles are far too complex, and perform far too many tasks, to not stimulate the human anthropomorphising tendency to some degree.

And in marketing circles the benefits of anthropomorphism towards increasing commercial sales have been routinely noted. For example, Hart and Royne (2017) investigated several forms of anthropomorphic representation and found that they all enhanced advertising effectiveness, brand awareness and purchasing intentions. Anthropomorphised products and anthropomorphic advertisements have been repeatedly observed to outperform those of competitors which did not stimulate the human tendency. Careful control of the anthropomorphising would thus also seem important towards commercial success.

Anthropomorphising Autonomous Road Vehicles

Automobiles have many components and characteristics which can stimulate the human anthropomorphising tendency. For example, the shape and colour of the headlights and taillights can stimulate anthropomorphic interpretations as eyes. And the use of a large grille at the front of the vehicle, even in cases where the airflow requirements of modern cooling systems no longer dictate a need for a large aperture, suggests the possible benefit of having some form of mouth at the front of the vehicle to complete its face.

The human anthropomorphising tendency has been noted to influence many aspects of automotive design and of car culture. For example, automobiles have often been named in an anthropomorphic manner in marketing campaigns, as historical examples such as the Austin Healey Frogeye Sprite of the late 1950s confirm. And the anthropomorphism provided by the headlights, grilles and bumpers served as a basis for many works of fiction such as the series of Walt Disney films which described the adventures of Herbie The Lovebug and his Lancia Montecarlo and other anthropomorphised friends.

With automobiles and other road vehicles there are also many opportunities for stimulating the human anthropomorphising tendency through motion. Headlight opening, window lifting, door opening, boot hatch lifting, radio antenna extraction and many other motions can stimulate associations with living creatures. Similarity in the type of motion, its spatial extent or its speed profile can lead to associations with familiar motions such as those of other living creatures or even those of humans.

And anthropomorphic responses can be stimulated by the acoustical and visual stimuli which the automobiles emit. For example, Dunne (2003) noted that sound quality surveys of Jaguar-branded automobiles frequently contained feline-related comments and that "To use the feline phrases associated with the Jaguar name, the requirement is to purr at idle and to growl when accelerating". The shaping of sound and light emissions has often been influenced by anthropomorphic reactions, and it is no secret that these reactions are routinely exploited for purposes of driving experience enhancement and branding.

And, of course, the infotainment systems of current automobiles and the passenger interfaces of many recent autonomous road vehicle prototypes involve voice interfaces which strongly stimulate the human anthropomorphic tendency. The semantics and grammatical structures which are chosen by the designers for these systems provide anthropomorphic cues. And design characteristics such as the voice's gender, speed of talking, tone and discussion domain can all stimulate the human anthropomorphising tendency and lead to the activation of cognitive biases, stereotypes and behaviour patterns.

Despite the novelty of autonomous road vehicles the last decade has already seen some research about these machines, and several studies have included some form of assessment of anthropomorphism. It has not escaped the attention of the designers that self-driving provides an illusion of intelligence and agency, even in the absence of aesthetic or behavioural supports to the human anthropomorphising tendency. A machine which adjusts real-time behaviours in response to the local contextual constraints cannot help but stimulate anthropomorphic thoughts and analogies.

While a variety of curiosities and questions have already been investigated, review of the studies performed to date suggests three obvious areas which are affected to some degree by the human anthropomorphising tendency. All three are made possible by the increase in complexity with respect to traditional, driven, 20th century automobiles. All three have been topics of research in the field of robotics, all three have been topics of philosophical speculation dating back many millennia, and all three are deeply entwined with the phenomenon of anthropomorphism. The obvious areas are: agency, trust and friendship.

The definition of "agency" which is most relevant to the design of autonomous road vehicles is usually stated in dictionaries as something along the lines of "the ability to take action or to choose what action to take". This sense of the word shares some attributes with words such as "ability", "power" and "freedom".

The issue of "agency" arises in relation to autonomous road vehicles because the vehicle is doing the driving much of the time. It was rare in the 20th century to speak of the "agency" of an automobile even when driving at speed or facing technically challenging road conditions. It is instead now increasingly common to hear people speak of the "agency" of autonomous road vehicles, even though the term may not be strictly appropriate technically or philosophically.

Nyholm (2018) has suggested four types of agency which can characterise any form of robot:

- domain-specific basic agency: pursuing goals on the basis of representations, within certain limited domains;
- domain-specific principled agency: pursuing goals on the basis of representations in a way that is regulated and constrained by certain rules or principles, within certain limited domains;
- domain-specific supervised and deferential principled agency: pursuing a goal on the basis of representations in a way that is regulated by certain rules or principles, while being supervised by some authority who can stop us or to whom control can be ceded, at least within certain limited domains;
- domain-specific responsible agency: pursuing goals in a way that is sensitive to representations of the environment and regulated by certain rules/principles for what to do/not to do (within certain limited domains), while having the ability to understand criticism of one's agency, along with the ability to defend or alter one's actions based on one's principles or principled criticism of one's agency.

Each type involves a somewhat different set of constraints and capabilities (Nyholm 2020) and the system of classification implies increasing complexity and responsibility when moving across the types from the first to the last. Naturalness of interaction, emotional intuitiveness and public acceptance of autonomous road vehicles may depend on the aesthetics, materials, functions, communication strategies and behaviours being consistent with the type of "agency" which is chosen for the vehicle.

The matter of "agency" is a growing divide between the passive road vehicles of the 20th century and the friendly neighbourhood robots of the 21st century. And if the act of self-driving alone were to prove sufficient to stimulate the anthropomorphising tendency in humans, independent of other vehicle characteristics, then anthropomorphism gets elevated higher up the list of design priorities.

It will not have escaped the attention of the reader that "agency" is a term used most often in relation to living creatures, and that "anthropomorphism" refers to the attribution of human traits, emotions, or intentions to objects or other living creatures. Anthropomorphism is a tendency to perceive human characteristics in the non-human, simplifying and rationalising what we see around us. The innate human anthropomorphising tendency is thus a tool which assists us in making sense and interpreting, helping to predict what might occur next. With highly complex machines such as autonomous road vehicles whose inner workings cannot be fully self-evident to their users it seems difficult to imagine that agency and anthropomorphism can be easily decoupled.

It may in fact come to light in the coming years that a minimum amount of autonomous road vehicle anthropomorphism is essential if it is to interact with people in a natural manner and transmit to them a sense of its capabilities and responsibilities. While driving, facilitating or entertaining, the autonomous road vehicle will be continuously selecting among the many preprogrammed actions associated with the form of agency chosen by the designers. And anthropomorphic cues and behaviours might be needed to communicate those actions and confirm the type of agency involved.

The second obvious area which has been researched and which is affected to some degree by the human anthropomorphising tendency is that of "trust". Many efficiency and safety issues arise when human operators do not trust the system which they are working with. The issue of "trust" arises in relation to autonomous road vehicles, again, because the vehicle is doing the driving much of the time. Do people enjoy being driven by someone or something which they do not "trust"? Whose abilities, professionalism or courtesy is suspect? Unlikely.

And, as with agency, for complex multifunction machines whose inner workings cannot be fully self-evident to their users it may prove challenging to decouple trust from anthropomorphism. Some anthropomorphic cues and behaviours might be needed to communicate decisions, clarify behaviours and build trust.

A simple machine such as a clock has a single visible output which is unlikely to be considered a behaviour. A complex multifunction machine such as an autonomous road vehicle has instead many moving parts and many service interactions which, when taken together, might be seen as following a pattern or exhibiting a behaviour. And the human anthropomorphising tendency pushes us to interpret such actions in terms of behaviours.

Among the autonomous road vehicle characteristics which can affect trust is the driving style. For example, Ekman et al. (2019) performed a series of Wizard Of Oz experiments which confirmed that driving style has obvious effects on passenger trust. Defensive driving styles were perceived as more trustworthy than aggressive driving styles, due in part to the greater degree of predictability. And since the human anthropomorphising tendency pushes us to see behaviours in the world around us, it may prove productive to design anthropomorphic cues and behaviours which communicate and clarify the driving style.

Among the autonomous road vehicle characteristics which can affect trust is also the manner of informing the passengers about the route and road events, and the manner of requesting the needed inputs from the passengers. Ruijten et al. (2018) have investigated the possible benefits of conversational interfaces over traditional visual displays for such purposes in autonomous road vehicles. Their findings suggested that conversational interfaces are anthropomorphised more and perceived as more intelligent, better liked and more trusted than traditional graphical user interfaces. Along similar lines, Forster et al. (2017) found that adding speech to an autonomous road vehicle's information system increased the degree of anthropomorphism and improved the trust and acceptance. The adoption of anthropomorphic cues and behaviours appeared helpful.

Confirming the role of anthropomorphism in relation to people's trust, Waytz et al. (2014) performed a driving simulator based study in which participants drove either a traditional automobile, an autonomous road vehicle which controlled the speed and steering, or an autonomous road vehicle which controlled speed and steering and which was further anthropomorphised by assigning it a name, gender and voice. Participant behaviour, physiological response and self-reported psychological metrics all suggested increases in trust in the vehicle as the anthropomorphic features increased. Again, the adoption of anthropomorphic cues and behaviours appeared helpful.

Attempting to draw some conclusions about trust in automation, Hoff and Bashir (2015) reviewed 127 research studies which were published between 2002 and 2013. Among their main conclusions was the suggestion that "In order to promote greater trust and discourage automation disuse, designers should consider increasing an automated system's degree of anthropomorphism, transparency, politeness, and ease-of-use".

The third obvious area which is affected to some degree by the human anthropomorphising tendency is that of "friendship". This issue can perhaps be considered to be a new concern in the case of road vehicles. While many studies have documented emotional ties between owners and their vehicles (Gossling 2017), the human-driven vehicles of the 20th century had insufficient automation to strongly stimulate feelings of shared values and reciprocity. However the new fully autonomous road vehicles, SAE level 5 to be precise, make decisions and perform acts which may be interpreted by some people in terms of friendship.

It is worth noting that recent years have seen the growing popularity of pet robots such as Aibo (Fujita 2001) which have become companions to thousands of people. More so than robots in factories or hospitals, pet robots are designed and produced specifically for the purpose of providing entertainment, companionship and even friendship. As these robots have repeatedly shown in practice, companionship and friendship do seem possible between robots and humans. If such relationships were to emerge with our future friendly neighbourhood robots then an additional set of technical, philosophical, ethical and legal questions will be raised for the designers to deal with, and for society more generally. And the importance of friendship in shaping human behaviour and supporting human happiness cannot be overstated.

Philosophical speculations about the nature of friendship have deep roots going back at least two and a half millennia. And across that time some considerations have been repeatedly discussed. For example, one frequently referenced categorisation was provided in the Nicomachean Ethics by Aristotle (Barnes 1984) who suggested three essential types of friendship:

- utility friendship: based on a specific benefit for one or both parties;
- pleasure friendship: based on a specific pleasure or fun for one or both parties;
- virtue friendship: based on good will, shared values, knowledge and mutual admiration.

Aristotle argued that the utility and the pleasure forms were imperfect. He claimed that virtue friendship was stronger, more meaningful, and the main goal when achieving the good life. Given its importance, virtue friendship was also claimed to produce greater displeasure and sense of betrayal when lost.

Whether the feelings which autonomous road vehicles stimulate in some people will be limited to "utility friendship" or instead be "pleasure friendship" or even "virtue friendship" is an open question. Design interventions involving targeted anthropomorphic characteristics, cues and behaviours may be needed to constrain the feelings which people develop in relation to the friendly neighbourhood robot. Deciding what form for friendship, if any, may turn out to be one of the major decisions facing the designers of our future friendly neighbourhood robots.

Danaher (2019) has extended the Aristotelian analysis by suggesting that virtue friendships require the following conditions:

- mutuality (i.e. shared values, interests, admiration and well-wishing between the friends);
- honesty/authenticity (the friends must present themselves to each other as they truly are and not be selective or manipulative in their self-presentation);
- equality (i.e. the parties must be on roughly equal footing; there cannot be a dominant or superior party);
- diversity of interactions (i.e. the parties must interact with one another in many different ways/domains of life, not just one or two).

The Danaher conditions provide further insights into the nature of friendship and may prove to be useful criteria for shaping the friendships between the friendly neighbourhood robots and their human users. For example, design interventions which reduce the scope for one or more of the Danaher conditions may help to reduce the likelihood of the friendly neighbourhood robot stimulating human feelings which reach the "virtue friendship" type.

And it is in fact Danaher's view (2019) that virtue friendships with robots, and by extension with autonomous road vehicles, are indeed possible. He has argued that science fiction has often explored the concept of friendship with alien beings which are dramatically different from humans and that the resulting public reactions have suggested that such friendships are a real possibility. Danaher's observations can be extended to suggest that mutuality, honesty, equality and diversity are not speciesist, thus neither is virtue friendship.

And if there should prove to be few logical or philosophical imped-iments to friendships with non-humans, then it will also prove to be the designers who will have to suggest the degree of friendship (utility, pleasure or virtue) which is consistent with the friendly neighbourhood robot. Targets for such matters may be needed early in the concept design phase to constrain the design process and to provide a reference for the nature and degree of anthropomorphism to adopt for the vehicle.

Anthropomorphism And Human Centred Design

From a design perspective, particularly a Human Centred Design per-spective, the issue of anthropomorphism looms large. That is why the topic is discussed relatively early in this book. Most people appear to interact with automated systems differently when the system is anthro-pomorphic to some degree.

A simple mechanical tool such as a spanner or a clock has only a small number of functions and affordances, which usually remain invariant over time. A simple, causal, mechanical device is usually highly predict-able, thus it does not strongly stimulate the human anthropomorphising tendency. Once a person understands what a screwdriver is for, nobody would expect the screwdriver to be capable of driving a car or of want-ing to do so.

If it were decided that a given autonomous road vehicle should pre-sent itself to the world as a simple mechanical tool, then the objective of the design team would necessarily be to achieve simple causal relation-ships everywhere possible. The team would wish to achieve only a small number of functions which are characterised by high levels of precision, accuracy and repetitiveness of operation. Achieving a high degree of predictability would be one way of reducing the impact of the human anthropomorphising tendency.

Such a machine might be constrained to transport people from point A to point B over fixed routes, possibly characterised by short distances and low speeds. The machine's ability to adapt to conditions would be limited by the need for simplicity, as probably would also be the degree of accessibility, universality and inclusivity achieved. Perhaps some cur-rent commercially available pods and robo-taxis might be considered to be members of such a category of autonomous road vehicle.

If instead it were decided that a given autonomous road vehicle should present itself to the world as a complex multifunction machine, then the situation becomes more challenging. Such a vehicle would provide several distinct capabilities which are engaged with via several distinct interactions. And the multiple capabilities might be executed differently in different contexts, and perhaps even at different times. Due to the complexity, it seems unlikely that such a vehicle would prove to be fully understandable and fully predictable by all of the people all of the time.

The complex multifunction machine would have motions and interactions which are not fully predictable but which, when taken together, might be interpreted in terms of behaviours. With such a machine it would be important to choose the degree of anthropomorphism and the exact anthropomorphising interactions, a priori, and to stick with them throughout the design process. The naturalness of interaction, emotional intuitiveness and public acceptance would likely depend on the aesthetics, materials, forces, motions, functions and communication strategies being consistent with the chosen degree of anthropomorphism. Many people might not be able to accurately predict the actions, but they might be able to accurately interpret the behaviours.

Not sticking to the constraints which are dictated by the intended degree of anthropomorphism could lead to confusion. For example, a friendly neighbourhood robot which is responsible for both the driving and the passenger safety might have difficulties conveying the urgency of an emergency situation if it were equipped with only a visual display for interacting with the passengers. Consider the importance of the acoustical vocal cues when an airplane pilot shouts out the order to adopt the crash position during an emergency. A substantial body of research describes the ways in which changes in sound pitch or word inflection affect the human interpretation of the urgency of a situation (see for example Stanton and Edworthy 2018). Thus the inability to produce acoustical or natural language instructions might make many communications difficult for the friendly neighbourhood robot.

A lack of a consistency in the stimulation of the human anthropomorphising tendency could therefore lead to confusion, and the doubts might not prove to be the same for everyone. The uncertainties would not be expected to manifest themselves in exactly the same ways with different people. Different people would react differently, thus further adding to the already impressive list of challenges in achieving accessibility, universality and inclusivity. It is already difficult enough to design a machine which is intended for use by a wide range of people without having to also deal with such further complications.

And it is also the case that the opportunities for business and public service development may depend on the degree by which the human anthropomorphising tendency is stimulated. Studies such as those of Rauschnabel and Ahuvia (2014) have for example suggested that brand love is best modelled on human interpersonal relationships. And the

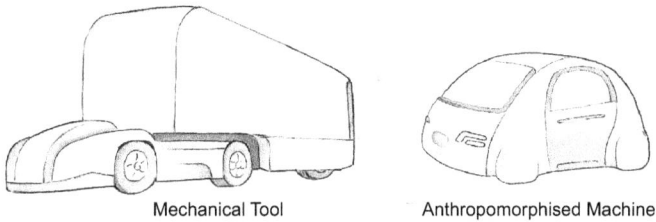

Figure 4.2 The choice between a simple mechanical tool or an anthropomorphised machine.

Source: Henry Leeson

more strongly the human anthropomorphising tendency is stimulated the more human and interpersonal the interactions with the machine may appear. The degree of anthropomorphising will thus likely influence how people talk about the friendly neighbourhood robot, how enamoured they become with the services which it offers, and how willingly they might accept it on their streets.

From a Human Centred Design point of view it would seem important that the first lines of the autonomous road vehicle design brief establish whether it is to present itself as a simple mechanical tool or instead as a complex multifunction anthropomorphised machine (see Figure 4.2). This one decision leads to a cascade of downstream functional, material, aesthetic and interaction requirements.

Conclusions

Anthropomorphism refers to the attribution of human traits, emotions, or intentions to non-human entities with a view to rationalising their actions. It is an innate human tendency which becomes increasingly more stimulated with increasing levels of behavioural complexity, and one which should be considered during autonomous road vehicle design. This chapter has provided a short introduction to the concept of anthropomorphism, to the anthropomorphising of animals, to the anthropomorphising of machines, to the anthropomorphising of autonomous road vehicles and to the relationship between anthropomorphism and Human Centred Design.

It was noted that the historical record suggests that humans have anthropomorphised their interactions with animals and their descriptions of them from early times. Evidence also exists for the anthropomorphising of machines, or at least for the anthropomorphising of those machines which were sufficiently multifunctional and complex to appear to human eyes to be exhibiting behaviours.

Among the more interesting manifestations of the human anthropomorphising tendency which were introduced in this chapter were the psychological concept of "uncanny valley" and the sociological phenomenon of the "Tamagotchi". The first provides a threshold beyond which the machine is no longer being considered by people to be only "anthropomorphic", but is actually now being considered "human-like". The second is a much studied phenomenon which reveals how people can develop strong ritualistic interactions with, and emotional attachments to, inanimate objects. Each reveals specific truths about how the anthropomorphising tendency is stimulated and how it affects people's interaction with machines.

This chapter also cited studies of anthropomorphism in relation to road vehicles. Aesthetics, materials, forces, motions, functions and communication strategies can all be used to stimulate the human anthropomorphising tendency if so desired. Much evidence suggests that people's interactions with road vehicles and their opinions about them tend to be influenced by the anthropomorphic characteristics and cues. This chapter also cited several studies which were specific to autonomous road vehicles and noted that the anthropomorphising of characteristics and cues strongly affects people's opinions about the vehicle in three key conceptual domains: agency, trust and friendship.

The issue of "agency" arises in relation to autonomous road vehicles because the vehicle is doing the driving much of the time. It was noted that four types of agency have been defined to characterise any form of robot: domain-specific basic agency, domain-specific principled agency, domain-specific supervised and deferential principled agency and domain-specific responsible agency. Each type involves a somewhat different set of constraints and capabilities, and the system of classification implies increasing complexity and responsibility when moving across the types from the first to the last.

The issue of "trust" arises in relation to autonomous road vehicles because, again, the vehicle is doing the driving much of the time. A simple machine such as a clock has a single visible output which is unlikely to be considered a behaviour. A complex multifunction machine such as an autonomous road vehicle has instead many moving parts and many service interactions which, when taken together, might be seen to be exhibiting a behaviour. It was noted that many research studies of trust in automation have found that a degree of anthropomorphism tends to support and facilitate trust.

The issue of "friendship" was noted to be new to automotive design. While 20th century road vehicles did not usually instigate such feelings in humans, the 21st century friendly neighbourhood robots may end up doing so. The Aristotelian categories of utility friendship, pleasure friendship and virtue friendship were introduced to illustrate the existence of different degrees of emotional attachment, and the Danaher conditions on virtue friendship were proposed as possible guidelines for constraining or limiting the development of friendship between people and their friendly neighbourhood robots.

Finally, this chapter noted the importance of deciding the degree of anthropomorphism and the exact anthropomorphising interactions, a priori, and sticking with them throughout. It was suggested that the aesthetics, materials, forces, motions, functions and communication strategies need to be consistent with the target level of anthropomorphism. With time, it is likely the public acceptance of the friendly neighbourhood robots will be shown to depend on such consistency.

Having introduced what is perhaps the biggest autonomous road vehicle design decision, that of reducing or instead amplifying the human anthropomorphising tendency, the next chapter will introduce a property which is a major consideration in relation to any artefact: its name.

References

Allison, A. 2006, Millennial Monsters: Japanese toys and the global imagination, Vol. 13, University of California Press, Berkeley, California, USA.

Bacon, F. (1960 [1620]), The New Organon And Related Writings. Prentice Hall, New York, New York, USA.

Barnes, J. ed. 1984, Complete Works Of Aristotle, Volume 2: The revised Oxford translation (Vol. 2), Princeton University Press, Princeton, New Jersey, USA.

Bartz, J.A., Tchalova, K. and Fenerci, C. 2016, Reminders Of Social Connection Can Attenuate Anthropomorphism: a replication and extension of Epley, Akalis, Waytz, and Cacioppo (2008), Psychological Science, Vol. 27, No. 12, pp. 1644–1650.

Benfield, J.A., Szlemko, W.J. and Bell, P.A. 2007, Driver Personality And Anthropomorphic Attributions Of Vehicle Personality Relate To Reported Aggressive Driving Tendencies, Personality And Individual Differences, Vol. 42, No. 2, pp. 247–258.

Bernotat, J., Eyssel, F. and Sachse, J. 2017, Shape It – the influence of robot body shape on gender perception, Ninth International Conference On Social Robotics, November 22nd–24th, Tsukuba, Japan, pp. 75–84.

Bliss, G. 2012, Animals With Attitude: finding a place for animated animals, Proceedings of the Conference: Critical Perspectives on Animals in Society, University of Exeter, UK, March 10th.

Campbell, G.L. ed. 2014, The Oxford Handbook of Animals in Classical Thought and Life, Oxford University Press, Oxford, UK.

Crist, E. 2010, Images Of Animals, Temple University Press, Philadelphia, Pennsylvania, USA.

Danaher, J. 2019, The Philosophical Case For Robot Friendship, Journal of Posthuman Studies, Vol. 3, No. 1, pp. 5–24.

Daston, L. and Mitman, G. eds. 2005, Thinking With Animals: new perspectives on anthropomorphism, Columbia University Press, New York, USA.

Duffy, B.R. 2003, Anthropomorphism And The Social Robot, Robotics And Autonomous Systems, Vol. 42, No. 3–4, pp. 177–190.

Dunne, G.T. 2003, The Introduction Of A Sound Quality Engineering Process To Jaguar Cars, Doctoral Dissertation, University of Warwick, Coventry, UK.

Ekman, F., Johansson, M., Bligård, L.O., Karlsson, M. and Strömberg, H. 2019, Exploring Automated Vehicle Driving Styles As A Source Of Trust Information, Transportation Research Part F: traffic psychology and behaviour, Vol. 65, pp. 268–279.

Forster, Y., Naujoks, F. and Neukum, A. 2017, Increasing anthropomorphism and trust in automated driving functions by adding speech output, IEEE Intelligent Vehicles Symposium (IV), June 11th to 14th, Redondo Beach, California, USA, pp. 365–372.

Fujita, M., 2001, AIBO: toward the era of digital creatures, The International Journal Of Robotics Research, Vol. 20, No. 10, pp. 781–794.

Gossling, S. 2017, The Psychology Of The Car: automobile admiration, attachment, and addiction, Elsevier, Amsterdam, Netherlands.

Hara, F. and Kobayashi, H. 1995, Use Of Face Robot For Human–Computer Communication, Proceedings Of The International Conference On Systems, Man and Cybernetics, Vancouver, British Columbia, Canada, October 22nd–25th, p. 10.

Hart, P. and Royne, M.B. 2017, Being Human: how anthropomorphic presentations can enhance advertising effectiveness, Journal Of Current Issues & Research In Advertising, Vol. 38, No. 2, pp. 129–145.

Hiroi, Y. and Ito, A. 2008, July. Are Bigger Robots Scary?—the relationship between robot size and psychological threat, IEEE/ASME International Conference On Advanced Intelligent Mechatronics, July 2nd–5th, Xian, China, pp. 546–551.

Hoff, K.A. and Bashir, M. 2015, Trust In Automation: integrating empirical evidence on factors that influence trust, Human Factors, Vol. 57, No. 3, pp. 407–434.

Hume, D. 1757, The Natural History Of Religion, Section 3.

Lawton, L. 2017, Taken By The Tamagotchi: how a toy changed the perspective on mobile technology, The iJournal: Graduate Student Journal of the Faculty of Information, Vol. 2, No. 2, pp. 1–8.

Maeng, A. and Aggarwal, P. 2018, Facing Dominance: anthropomorphism and the effect of product face ratio on consumer preference, Journal Of Consumer Research, Vol. 44, No. 5, pp. 1104–1122.

Miesler, L. 2011, Imitating Human Forms In Product Design: how does anthropomorphism work, when does it work, and what does it affect, PhD Dissertation, The University of St. Gallen, St. Gallen, Switzerland.

Mithen, S.J. 1998, The Prehistory Of The Mind: a search for the origins of art, religion and science, Phoenix Paperback, London, UK.

Mori, M. 1982, The Buddha In The Robot, Charles E. Tuttle Co., North Clarendon, Vermont, USA.

Nomura, T. 2017, Robots And Gender, Gender And The Genome, Vol. 1, No. 1, pp. 18–26.

Nyholm, S. 2018, Attributing Agency To Automated Systems: reflections on human–robot collaborations and responsibility-loci, Science And Engineering Ethics, Vol. 24, No. 4, pp. 1201–1219.

Nyholm, S. 2020, Human And Robots: ethics, agency and anthropomorphism, Rowman & Littlefield, London, UK.

Pazhoohi, F. and Kingstone, A. 2020, Larger Vehicles Are Perceived As More Aggressive, Angry, Dominant, And Masculine, Current Psychology, pp. 1–5.

Rauschnabel, P.A. and Ahuvia, A.C. 2014, You're So Lovable: anthropomorphism and brand love, Journal Of Brand Management, Vol. 21, No. 5, pp. 372–395.

Reeves, B. and Nass, C. 1996, The Media Equation: how people treat computers, television, and new media like real people, Cambridge University Press, Cambridge, UK.

Ruijten, P.A., Terken, J. and Chandramouli, S.N. 2018, Enhancing Trust In Autonomous Vehicles Through Intelligent User Interfaces That Mimic Human Behaviour, Multimodal Technologies And Interaction, Vol. 2, No. 4, p. 62.

Stanton, N.A. and Edworthy, J. 2018, Human Factors In Auditory Warnings, Routledge, Abingdon, Oxfordshire, UK.

Temple, R. 2003, Aesop – The Complete Fables, Penguin Books, London, UK.

Turner, M. 1987, Death Is The Mother Of Beauty: mind, metaphor, criticism, University of Chicago Press, Chicago, Illinois, USA.

Waytz, A., Morewedge, C.K., Epley, N., Monteleone, G., Gao, J.H. and Cacioppo, J.T. 2010, Making Sense By Making Sentient: effectance motivation increases anthropomorphism, Journal Of Personality And Social Psychology, Vol. 99, No. 3, pp. 410–435.

Waytz, A., Heafner, J. and Epley, N. 2014, The Mind In The Machine: anthropomorphism increases trust in an autonomous vehicle, Journal Of Experimental Social Psychology, Vol. 52, pp. 113–117.

Wodehouse, A., Brisco, R., Broussard, E. and Duffy, A. 2018, Pareidolia: characterising facial anthropomorphism and its implications for product design, Journal Of Design Research, Vol. 16, No. 2, pp. 83–98.

Wu, L., Larkin, M., Potnuru, A. and Tadesse, Y. 2016, HBS-1: a modular child-size 3D printed humanoid, Robotics, Vol. 5, No.1, p. 1.

Chapter 5

Name

Name

Dictionaries suggest that a "name" is a word or group of words by which a thing or person is known, addressed, or referred to. Names are usually thought of as nouns, i.e. words which refer to something and which carry connotations and associations. Names are at the core of communication and language.

In *The Theory Of Moral Sentiments* Adam Smith (1759) wrote that "the assignation of particular names to denote particular objects, that is, the institution of nouns substantive, would probably be one of the first steps towards the formation of language. Two savages who had never been taught to speak, but who had been bred up remote from the societies of men, would naturally begin to form that language by which they would endeavour to make their mutual wants intelligible to each other, by uttering certain sounds, whenever they mean to denote certain objects."

Mollerup (2013) suggested that both names and numbers are tools for thinking, and that descriptive names are more useful than simple designations. And Chandler (2003) has emphasised that "the communicative function of a fully-functioning language requires the scope of reference to move beyond the particularity of the individual instance. While each leaf, cloud or smile is different from all others, effective communication requires general categories or universals".

Approaching names and naming from a sociological perspective, Charmaz (2006) wrote: "As symbolic interactionists have long argued, names classify objects and events and convey meanings and distinctions. Names carry weight, whether light or heavy. Names provide ways of knowing and being. Names construct and reify human bonds and social divisions. We attach value to some names and dismiss others."

The names which we assign to things or to people are thus much more than a simple pointer. They direct, identify and connect. They are

DOI: 10.4324/9781003319740-5

complex systems of connotation and association which serve as a basis for thinking and for communicating. They are functionally necessary and culturally significant in most human activities.

Scholars have debated the nature of names for centuries and these days two streams of thought seem to dominate. The first can be called the Description Theory Of Names while the second is often referred to as the Usage Theory Of Names (Evans and Altham 1973). While there are several differences between the two approaches, the main one lies in what should be considered the constitutive source of the name.

The first view assumes that a name refers to something which exists independently and whose attributes are knowable a priori. It suggests a process in which people discover the appropriate names for the things around them based on the essences of those things, and on the possibilities for matching those essences with sounds which are vocalisable by humans. The second view tends instead to focus on how a name becomes constituted through its use by people within conversations within a society. The second view considers names to be more arbitrary in nature, and more subject to outside influences which may have affected the naming at an early stage. As a name may sometimes be more the result of social forces than of any specific characteristics of the artefact or person involved, the second view also implies the benefit of articulating a few key attributes along with the name itself, so as to ensure that the full spectrum of connotations and associations is stimulated. This second, more constructivist, view is of particular interest in this chapter.

Any analysis of naming (see for example Hough and Izdebska 2016) is complicated by the fact that names can simultaneously denote and connote. A name can be highly specific, perhaps unique to a single artefact or single individual, acting to denote that one example. However, names can also be more general, implying a set of characteristics by connotation. While a "Fiat 500" might be a specific automobile with specific characteristics, the "Fiat" component of the name may bring with it connotations such as "small", "fuel efficient", "Italian" and "nostalgic". Parts of a name, or the complete name, can connote and associate well beyond the individual example. In fact, corporate names such as "Fiat" are often more important in terms of the brand values which they imply than the individual artefacts which they denote.

But any analysis of the effects of naming suggests that distinctive names can bring important functional, economic and societal benefits. And one approach to finding such names was first suggested by Hartmann (1966) who noted that information theory (Shannon 1949) can numerically quantify the distinctiveness of a word or phrase. Information theory posits that the amount of information provided by a word or by a phrase is proportional to its rarity of usage in the given language. Information theory calculations are based on the computed frequency of occurrence of the given word or phrase in one or more major linguistic databases. A rather uncommon word such as "supercalifragilisticexpialidocious" is thought to provide more information than a common word such as "the". There is greater novelty and less room for error in identification when the name is rare and unusual. Using letters, words or phrases which are rarely encountered in everyday life usually leads to those names proving more attention grabbing.

And information theory calculations tend to support everyday experiences, such as finding the letter "x" in the names of many new products, systems and services. The letter "x" is of course among those which are statistically encountered the least in everyday conversation, thus its use attracts attention. A quick look around suggests that it is not difficult to find examples of names which involve unusual letters from other alphabets, rarely used letters, unusual letter combinations or complete words which are unexpected in relation to the artefact. Many commercial names are creative, rare, attention grabbing and high in information.

And beyond the amount of information provided by a name there is also the matter of its nature and meaning. The connotations and associations which define any given name form a web of connectivities within which the name resides. Techniques such as protocol analysis (Ericsson and Simon 1984) and content analysis (Krippendorff 2004) are based on the use of those connectivities for making sense of cyphers and messages. Calculating the degree of correlation between the individual words and names which are contained in a message, or within an ensemble of messages, can provide clues as to the underlying structure and meaning of the communication. When words or names are routinely found together it suggests some connection, assumed or real, in the mind of the message's writer. The choice of any given name is thus as much a matter of selecting its web of connectivities as it is one of denoting a specific person or artefact. Consider for example meeting two boys over the course of a day, one named Alessandro and the other named Hiroshi, what might come to mind in each case?

Names help people make sense of things and of situations. And as needs change, they evolve along with the people and the society. Names can be associated with a specific geographical area or a specific historical period. They can also be typical of some technology or of some culture. And they can indicate as much by their absence as by their presence. Consider for example what the absence of a formally assigned name might say about the importance of some product, or what it might mean for someone's name to be left off a work roster or a list of authors.

From a neuroscience perspective it has been repeatedly noted that memory recall is better for names which are used more frequently (see for example Bredart in Hough and Izdebska 2016) than for those which are encountered infrequently. Neuronal storage, via changes in synaptic density and neurotransmitter release propensities, becomes stronger and more complete as the number of exposures increases. And as any educator knows, the accuracy of recall of a word, sentence or physical movement improves with increasing number of repetitions up to some maximum. Familiarity with a name, or lack of it, is thus a non-trivial consideration. Will a proposed name elicit the appropriate connotations and associations in people's minds? If not, can such connotations and associations be established over time?

Naming Animals

It has been suggested (see Leibring in Hough and Izdebska 2016) that humans probably began naming animals as soon as the first animals were tamed. It is certainly the case that archaeologists encountered named animals from the Egyptian Old Kingdom and that there are clay tablets from Crete written in Linear B which refer to specific oxen using names such as "Darkie" or "Whitefoot".

According to Leibring "not all animals living close to humans are given an individual name, but there are several requisites that can make naming more probable". Conditions suggested to stimulate naming included:

- the animal is expected to live for several years and is not regarded as food;
- the animal is in some way distinctive in its outer appearance;
- the animal is one of several in a flock or herd;
- there is a need for identification of the animal or for communication with it;
- the animal is considered or treated as an individual by the handler or owner.

Researchers usually draw a distinction between the naming systems used for production/utility animals such as cows and sheep, and those used instead with companion animals such as cats and dogs. Historical analysis has suggested that animal names have tended to depend on the perceived degree of functional utility, ritualistic value or emotional/mythical value assigned to the animal. The category of meaning assigned to the animal or group of animals seems to have shaped the naming processes.

In the case of production/utility animals, researchers (see for example Leibring in Hough and Izdebska 2016) have suggested that the continuous contact between farmers and their animals in pre-mechanical eras usually led to names which were based on the physical or behavioural characteristics of the animal (such as its colour or temperament) or on the time or place of its birth or purchase. When animals were few, functional individual names appear to have been the norm.

As farm sizes and herd sizes grew the naming processes adapted to the new contexts. Researchers have suggested that from the 18th century onwards the individual animals of a herd were usually referred to according to the characteristics of the mother animal. The herd as a whole, or individual members of the herd, would be referred to in terms of either the mother animal's physical characteristics, or its productive characteristics, or its place of birth, or year of birth or geographical region of birth.

Addressing the needs of today, production/utility animals are usually catalogued in the European Union via officially assigned numbers which identify the individual animal, the source farm and the country of origin. Such numbers provide a system which is functional towards tracking and certifying individual animals, but is of course far removed from the characteristics of the individual animal or its life on a specific farm.

For the naming of companion animals there is evidence dating as far back as the Egyptian Old Kingdom (Fischer 1977) of the use of characteristics such as colour or temperament, or the animal's place of origin. Physical, behavioural and geographical characteristics seem to also have been common in early companion animal names.

An abundance of sources from classical antiquity contain companion animal names which are based on physical or behavioural characteristics, or on geographical origin. The historian Xenophon (430 to 354 BCE) in the "Cynegeticus" listed common dog names of the time including Lance, Lurcher, Watch, Keeper, Butcher, Craftsman, Forester, Counsellor, Spoiler, Hurry, Fury, Growler, Rome, Hebe, Hilary, Gazer, Eyebright, Force, Trooper, Bubbler, Rockdove, Stubborn, Yelp, Killer, Sky, Sunbeam, Wistful, Tracks and Dash, among others. It was also Xenophon who provided the first recorded evidence of the suggestion that dog names should be short, such that they can be quickly spoken and easily understood by the animal.

Roman and medieval dog names were more varied and were sometimes the personal names of divinities or those of well-known historical figures. And in the present day the naming of a family dog can be somewhat arbitrary, involving any characteristic or association which is meaningful to the people involved. An assortment of options is often considered and the name which is finally assigned to the animal can prove meaningful within the context, but not always fully interpretable outside the immediate family or setting. In today's pluralistic and individualistic societies the naming of the loved family pet can be another opportunity for creativity and for personal expression, leading to an ever expanding horizon for what can or cannot serve as an animal's name.

And, finally, it should be mentioned that studies of cat naming have suggested similarities to the naming of dogs. While cats were domesticated much more recently than dogs, leading to a smaller number of historical sources, those which are available contain many examples of physical, behavioural and geographical characteristics being used. Leibring (see contribution in Hough and Izdebska 2016) has also noted a small number of peculiarities which seem specific to cat naming, including the use of names which are based on the sounds (meowing, purring, etc.) which the animal makes. Also peculiar to cat naming is a greater phonological variety among the names, which Leibring suggests may be due to the reduced functional constraints. Cats do not always respond to the calling of their owners, thus short, distinctive functional names are not necessarily helpful in practice.

Naming Machines

Beyond animals, another important historical exercise in naming has been that of machines. Human history seems characterised by a progressive shift from an early dependence on animals for work and play to an increasing reliance on machines for those same purposes. And in the same way that naming was helpful in everyday interactions with animals, the naming of machines has also provided opportunities for simplification and efficiency. Autonomous road vehicles are not the first, nor will they be the last, machines which can benefit from a well-chosen name.

And for much of the last two centuries the archetypal machine in the public imagination has probably been the locomotive, or, more generally, the train. The naming of locomotives is rich in historical insights and provides evidence of the societal processes involved when some artefact is introduced, for better or for worse, into a human society.

Coates (see Hough and Izdebska 2016) has reviewed the names of British railway locomotives and has found that those of pre-1846 locomotives fell mostly into five categories:

- new: such as "Experiment", "Pioneer" and "Steam Horse";
- powerful: such as "Dangerous", "Lion" and "Vulcan";

- quick: such as "Catch Me Who Can", "Dart" and "Lightning";
- mechanical: such as "Puffing Billy" and "Steam Elephant";
- excellent: such as "Success", "Perseverance" and "Victoria".

The five categories are obviously descriptive in nature, providing a functional lens through which to view the artefact. Novelty appears to have driven the early naming of locomotives with the new and salient characteristics providing opportunities for explaining, promoting and advertising. Such an approach perhaps assists the societal assimilation of the new technology since essential characteristics of the machine are brought to mind, and highlighted, via the name.

Coates further suggested that after approximately 1846 a change began to occur in which the early functional naming of locomotives gave way to more commemorative names based on places, persons or events. Locomotive naming began to search beyond the machine itself for inspiration, and the growing acceptance of the machine suggested opportunities for positive associations with other societal interests. Commemorative names such as "Caledonian" or "Lord Wellington" were popular for almost a century until naming became, in turn, mainly a matter of public relations and branding.

From the mid-20th century onwards a direct connection between a given locomotive on the one hand, and either its characteristics or some point of commemoration on the other, was not always obvious. Locomotive names have increasingly been based on sources and inspirations of varied nature, ranging from a key moment in a company's history to themes taken from the arts or from popular culture.

Coates' observations support the view that the historical trajectory of British locomotive naming followed a path which started with an emphasis on functional characteristics, then moved on to greater focus on commemorative and ritualistic potential, finally terminating in the modern-day tendency to adopt names from any narrative or myth which can serve to further the commercial success. It might be said that the early names helped explain while the later names were more intended to attract attention.

And the analysis of a variety of products, systems, services and social phenomena beyond locomotives led Coates to further propose a "general law of name development" which has been claimed to be widely applicable. In this view, the need to communicate and to explain the nature of a new technology leads to early naming which incorporates elements of the technical system or its key metaphors. Early naming tends to be based on what Coates refers to as "essential characteristics". As the technology becomes more familiar, however, and its function and benefits become less mysterious, the naming process begins to exploit opportunities for commemorative associations between the machine and persons, places, events or values which are judged positively by society. Such associations locate the machine within a web of connectivities which help to bring positive thoughts and feelings to mind. Eventually, as novelty and variety wane, nearly arbitrary commercial branding becomes the norm with the selecting of names being based mostly on social, media and commercial convenience. Eventually, as the technology becomes fully integrated into everyday life, the objective of the naming becomes simply one of attracting attention.

As in the case of British locomotives it has also been noted that a large number of automobile names used by American manufacturers in the 19th and 20th centuries were based on "essential characteristics". An analysis by Piller (1996) of 2241 names of American automobiles produced between 1805 and 1996 noted that the "essential characteristics" could be grouped into a small number of metonyms and metaphors.

Both metonyms and metaphors are ways of describing something by means of reference to something else which is familiar and often simpler. Metonyms involve greater spatial, temporal or causal interrelatedness than do metaphors, but both are ways of helping to understand something by referring to something else. In language, both metonyms and metaphors are extensively used in everyday life and many researchers (see for example Lakoff 2008) even suggest that metonym and metaphor are the basis for human thoughts.

Piller (1996) noted the metonyms:

- the places where the car might be used (New Yorker, etc.);
- the purposes which the car might serve (off-road, racing, etc.);
- the producer of the car or of selected parts of the car (Dodge, Ford, etc.);
- the technical characteristics of the car (such as engine size);
- the special interest parts of the car (such as Shelby engine);
- the model, line or edition of the car (such as economy, deluxe, etc.);

and the metaphors:

- an element or phenomenon of unanimated nature (such as tornado);
- a man-made object other than a car (such as rocket);
- an animal (such as mustang);
- a human being (such as Edsel);
- a supernatural being (such as demon).

And, again as in the case of the British locomotives, it was noted that a large number of the automobile names used by American manufacturers in the late 20th century were more commemorative and associative in nature. While not every American automobile name can be said to fit the simple pattern, it is nevertheless the case that the frequency of use of "essential characteristics" appears to have been in steady decline from the mid-20th century onwards.

Today's automobile names are inspired by a variety of sources which help to further the manufacturer's media footprint and commercial interests. Unusual letter or number combinations are common, misspellings are frequent and cultural referencing is popular. Commercial convenience appears to be the main criteria. Attracting attention seems to be the goal.

In fact, today, many automobile names can prove somewhat confusing in terms of what they are actually referring to technically. The number "500" in the name "Fiat 500" no longer refers to the cubic centimetres of the engine combustion chambers, a Mercedes-badged automobile may actually contain a Renault-designed engine, and an expensive electronic "self-driving package" may be capable of little more than motorway lane-changing. It is not difficult to sense that reference to "essential characteristics" is no longer essential for the commercial success of traditional human-driven automobiles.

The final case of machine naming which will be discussed here is that of robots. The arrival of substantial numbers of robots starting from the late 20th century onwards has placed many people into direct contact with machines which exhibit greater complexity of function and greater potential for social interaction than did the earlier technologies of locomotive or automobile. Over the last half century the field of robotics has produced many machines which perform multistep tasks which blur the lines between repetitive mechanical action and complex animal-like behaviour. And each has a name. And like the machines themselves, the names often seem to blur the lines between the mechanical and the lifelike.

Keay (2012) analysed the names found on the websites of more than 100 robot competitions of various type, and one trend which was noted was gender bias. For social agent robots, 26% of the names were found to be female in nature while for robot vehicles only 8% of the names were female in nature. Intentionally or unintentionally, the robot names appeared to reflect societal stereotypes and biases in relation to gender and occupation.

Another interesting trend noted by Keay among the competing robots was that more than two-thirds of the names exhibited biomorphic or lifelike attributes as opposed to simple mechanical attributes. The robot names tended to be less functional and more behavioural in nature, intentionally or unintentionally reflecting the complex and lifelike behaviour which they exhibited.

A final feature of Keay's database of robot names was that of hybridity. Many robot names were mechanical in nature such as Bit, Motbot and Robomatic XI and many others were anthropomorphic in nature, such as Athena, Betty and Ziggy. But there were also names such as Ada 1852, Cyber Ty, Johnny 5 and Talk-Bot which can be described as hybrids, i.e. which contain both a mechanical and an anthropomorphic element within the single name. As Keay and Graduand (2011) themselves noted, "Robot names show a high level of alive/not alive hybridity, rather than traditional human or pet naming practices".

Regarding hybridity, it can be speculated that some designers may choose such names as a way of clarifying that the "essential characteristics" of the robot include both the mechanical and the lifelike. Hybrid names seem to blur the line between the mechanical and the lifelike in a similar manner to many of the machines themselves. Presumably, many designers are approaching the naming task in the spirit suggested by Keay (2012) who wrote that "Naming a robot categorizes it, creates expectations and triggers social responses".

Given the categorising, expectation setting and social response priming it is not surprising that the robot names which are found in catalogues, scientific publications and literary works of science fiction should span a wide range of intellectual spaces. Today's galaxy of robots fulfil a galaxy of roles, thus their names likewise form a galaxy of denotations and connotations.

Given the complexity, one specialist agency which focusses on the naming of robots and of other forms of automation (Namerobot 2021) has suggested a set of categories which can help in interpretation. The categories were determined from analysis of a wide range of materials and publications, and provide a structural cross section of today's galaxy of robot names. The robot name categories include:

- letter/number combinations such as "T-800" of Terminator;
- acronyms such as "D.A.V.E." (Digitally Advanced Villain Emulator) of Batman;
- descriptions such as "Commander Data" of Star Trek The Next Generation;
- robot-specific syllables such as "Robo" or "Bot";
- abstract concepts such as the "Curiosity" rover;
- mythological names such as with the "Hydra" chess computer;

- fantasy names such as "Gort" of The Day The Earth Stood Still;
- plays on words such as "Anne Droid" of Doctor Who;
- human proper names such as "Rosie" of the Jetsons.

Naming Autonomous Road Vehicles

Autonomous road vehicles are a relatively recent development involving relatively new technologies. Initially, for most people, an autonomous road vehicle will be a blank sheet of paper. Past memories will not necessarily assist interactions or understanding all that much. And helpful metonyms and metaphors may not be immediately available. Upon first contact with a friendly neighbourhood robot most people will probably be grasping for any information which can assist in understanding the alien presence in their midst. And a name, particularly a simple and intuitive one, will be one of the few available items of information.

The histories of animal naming and of machine naming suggest a "general law of name development". Even in the absence of such historical evidence, common sense might suggest that it could be helpful to be somewhat descriptive when naming something new and unfamiliar. Whether the current internet-driven information explosion, with its immediacy of circulation and opinion forming, will change the historical pattern is an open question.

It does not appear unreasonable to think that there may be less need today for descriptive names which highlight essential characteristics. Humans now have nearly instantaneous access to quantities of information which were unimaginable in centuries past. And the rapidity with which people, artefacts and ideas come and go is neatly summarised by the well-known phrase from Andy Warhol's 1968 brochure: "In the future, everyone will be world-famous for 15 minutes".

Today, regardless of its complexity, does a new machine really require essential elements naming in order to be quickly understood and easily assimilated by society? Are the essential elements as relevant today to human understanding and societal acceptance as they were in the past? And are the elements which can be considered "essential" the same as in the past? Only time will tell.

One thing that can be said is that from a Human Centred Design perspective the intended meaning is the most important aspect of any design. Meaning is the concept which is at the top of the HCD triangle. It is the main objective. It is the goal. The achievement of a well-defined meaning, at least for the majority of the intended customers or constituency members, is thus essential. And an artefact's name, like its other characteristics, can support and contribute to that meaning.

In the previous chapter it was noted that a Human Centred Design point of view suggests the importance of clarifying whether the autonomous road vehicle presents itself as a simple mechanical tool or instead as a complex multifunction anthropomorphised machine. The intended nature of the machine constrains the meanings which it can occupy in people's lives. This single decision leads to a cascade of downstream requirements, which include the denotations and connotations of the autonomous road vehicle's name.

If the designer intends that the robot be viewed as a simple mechanical tool, then Darling's (2015) observation about the effect of the name on the human anthropomorphising tendency might be worth considering. In Darling's words: "Focusing also on framing by objectifying robots in language ('it') and encouraging names such as 'MX model 96283' instead of 'Spot' will probably not make anthropomorphism disappear completely, but it may have a helpful effect". The designer can help people to interpret the autonomous road vehicle as a simple mechanical tool by choosing a simple mechanical name which alludes to the vehicle's main function or main service provision. Words such as "box", "connector", "cube", "mover", "pod", "shuttle" or "transporter" might prove helpful elements of such a name.

If instead the designer intends that the robot be viewed as a complex multifunction anthropomorphised machine, then it may prove useful to consider instead Darling's (2015) observations about "hospital staff being friendlier towards robots that have been given human names" or those about participants who "hesitated significantly more to strike the robot when it was introduced through anthropomorphic framing (such as a name or backstory)". The designer can help people to interpret the autonomous road vehicle as a complex multifunction anthropomorphised machine by choosing a more complex and lifelike name which alludes to the vehicle's versatility, decision-making capabilities, responsibility and trust. Words such as "agent", "assistant", "doctor", "guide", "entertainer", "specialist" or "valet" might prove helpful elements of such a name.

The importance of the name given to a friendly neighbourhood robot cannot be overemphasised. For example, Darling (2015) has reported that hospital staff were friendlier towards a medicine delivery robot when it was given a human name as opposed to a numerical or non-human name. Even the tolerance for malfunction was higher. Statements such as "Oh, Betsy made a mistake!" were common when the robot was introduced using a human name, while statements such as "This stupid machine doesn't work!" were more common when the robot was introduced using a non-human name.

And a study performed by Eyssel and Kuchenbrandt (2012) suggested that, like people, robots can end up being assigned to a social group. In their study a robot was introduced differently to different groups of participants. The robot name and the clues to its provenance were introduced as being German for some participants and Turkish for others. Later, it was noted that cultural stereotypes were being applied to the robot in much the same manner that they might have been with a human of the same nationality.

The importance of any name lies in the position which the word or phrase occupies in the web of connotations and associations which it stimulates. Over the course of life each human being encounters a multitude of words and phrases in a bewildering range of contexts. And with each encounter, several key characteristics of the encounter are committed to memory alongside the word or phrase itself. Over time, words, phrases and environmental stimuli come to be deposited in human memory, forming patterns which, once stimulated for recall, lead to multiple simultaneous sensations and thoughts. Names thus carry positive or negative connotations and associations. Names are rarely neutral.

An illustrative example is provided by the study by Newman et al. (2018) in which 500 college students were asked to rate 400 popular male and female names. The questions adopted in the study were of the type: "Imagine that you are about to meet Samantha. How competent/warm/old do you think she is when you see her name?" The results for competency and warmth were:

- warm and competent names: Ann, Anna, Caroline, Daniel, David, Elizabeth, Emily, Emma, Evelyn, Felicia, Grace, James, Jennifer, John, Jonathan, Julie, Kathleen, Madeline, Mark, Mary, Matthew, Michael, Michelle, Natalie, Nicholas, Noah, Olivia, Paul, Rachel, Samantha, Sarah, Sophia, Stephen, Susan, Thomas, William;
- warm but less competent names: Hailey, Hannah, Jesse, Kellie, Melody, Mia;
- competent but less warm names: Arnold, Gerard, Herbert, Howard, Lawrence, Norman, Reginald, Stuart;
- names of low warmth and competence: Alvin, Brent, Bryce, Cheyenne, Colby, Crystal, Dana, Darrell, Devon, Dominic, Dominique, Duane, Erin, Larry, Leslie, Lonnie, Malachi, Marcia, Marco, Mercedes, Omar, Regina, Rex, Roy, Tracy, Trenton, Vicki, Whitney.

As this study and many others suggest, something as individual as a person's name brings with it a variety of connotations and associations which do not initially depend on the person involved. With time, connotations and associations which are specific to the individual can develop, but, initially, the name brings its own. As any parent knows, the choosing of a name for a newborn baby (Copeland 2017) involves navigating a minefield of possible meanings.

And the connotations and associations do not just come with a person's name but also with a person's surname. Several formal dictionaries (see for example Hanks et al. 2016 or Parkin 2021) have in fact been developed to document the historical sources and most obvious meanings of local surnames. Speaking specifically about proper names Rymes (1999) wrote, "From an anthropological perspective, names are not simply arbitrary labels. How we get them, who says them, how they are used, and in what context they are spoken are inseparable from a human being's social identity".

As the research about names suggests, patterns of connotation and association are stimulated every time one is used. And the activated patterns can feel intuitive, or counterintuitive, to the people involved for a variety of reasons which are not all fully accessible to rational conscious thought. How, where and why that name was encountered by the person in the past can have a big influence on how the person reacts in the present. While the first contact with an autonomous road vehicle may prove to be a complete novelty for many people, the name component of the contact may not be.

It may therefore prove unwise to choose the name of a friendly neighbourhood robot or of its service provision based only on creativity, novelty and uniqueness. Careful empirical exploration would seem beneficial given the complexity of the elicited patterns of connotation and association. Of the names provided alongside the autonomous vehicles which are shown in Figure 5.1 (below) and Figure 5.2 (on the following page), which of them feels more intuitive? And is it obvious why?

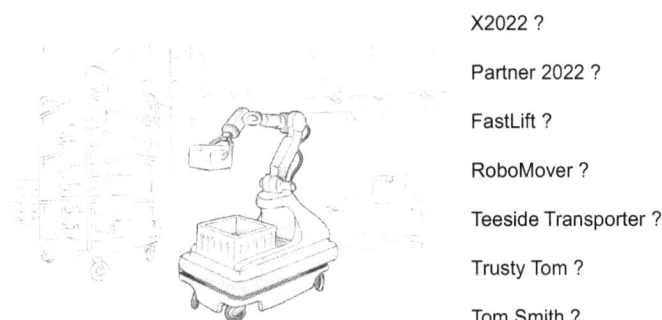

X2022 ?

Partner 2022 ?

FastLift ?

RoboMover ?

Teeside Transporter ?

Trusty Tom ?

Tom Smith ?

Figure 5.1 Naming a mechanical machine.

Source: Henry Leeson

X2022 ?

Partner 2022 ?

FastLift ?

RoboMover ?

Teeside Transporter ?

Trusty Tom ?

Tom Smith ?

Figure 5.2 Naming a complex multifunction anthropomorphised machine.

Source: Henry Leeson

Conclusions

Names are words or groups of words by which a thing or person is known, addressed, or referred to. Names categorise, create expectations and trigger social responses. They carry a galaxy of connotations and associations with them. This chapter has provided a short introduction to the concept of name, to the historical naming of animals, to the historical naming of machines and to the naming of autonomous road vehicles.

Several studies of naming were cited in relation to animals, locomotives, automobiles and robots. It was noted that there is historical evidence in support of a "general law of name development" in which the early naming of something tends to be based on its "essential characteristics", then, with time, naming becomes more commemorative or opportunistic in nature. The "general law of name development" may describe a human tendency to adopt descriptive and functional names with something new, so as to assist people and society to become more familiar with the development.

For the most complex type of machine which was reviewed, the robot, it was suggested that the names involve a wider variety of attributes than those of animals, locomotives or automobiles. The arrival of robots placed people into contact with machines whose functions can blur the lines between repetitive mechanical action and diverse animal-like behaviour, and like the machines themselves the names also appear to blur those lines. The characteristic of "hybridity", which involves a mixture of mechanical attributes and lifelike or humanoid attributes, was noted as being particularly common in robot naming.

The material of this chapter repeatedly noted that patterns of connotation and association are stimulated every time a name is used, and that those patterns can have positive or negative effects on the relationships which are established. Names are rarely neutral. It was suggested that if the designer intends that the robot be viewed as a simple mechanical tool, then words such as "box", "connector", "cube", "mover", "pod", "shuttle" or "transporter" might prove helpful elements of the name. It was also suggested that if the designer instead intends that the robot be viewed as a complex multifunction anthropomorphised machine, then words such as "agent", "assistant", "doctor", "guide", "entertainer", "specialist" or "valet" might prove helpful elements of the name.

Finally it was noted that spending time and effort to identify a helpful name would seem a good investment since it can get relationships with the friendly neighbourhood robot off to a good start. Names carry positive or negative connotations and associations. They are rarely neutral. It therefore does not seem excessive to suggest that the names which we give to our future friendly neighbourhood robots will greatly influence their initial contact with people, their core meaning and their eventual place in society.

Having introduced naming in relation to autonomous road vehicles the next chapter introduces something which is closely associated with the friendly neighbourhood robot's name: its meaning.

References

Chandler, D. 2003, Semiotics: the basics, Routledge, New York, New York, USA.

Charmaz, K. 2006, The Power Of Names, Journal Of Contemporary Ethnography, Vol. 35, No. 4, pp. 396–399.

Copeland, T. 2017, Baby Names: 12,000+ baby name meanings & origins, CreateSpace Independent Publishing Platform.

Darling, K. 2015, "Who's Johnny?" Anthropomorphic Framing in Human–Robot Interaction, Integration, and Policy, ROBOT ETHICS, Vol. 2, March 23rd.

Ericsson, K.A. and Simon, H.A. 1984, Protocol Analysis: verbal reports as data, The MIT Press, Cambridge, Massachusetts, USA.

Evans, G. and Altham, J.E.J. 1973, The Causal Theory Of Names, Proceedings Of The Aristotelian Society, Supplementary Vol. 47, pp. 187–225.

Eyssel, F. and Kuchenbrandt, D. 2012, Social Categorization Of Social Robots: anthropomorphism as a function of robot group membership, British Journal Of Social Psychology, Vol. 51, No. 4, pp. 724–731.

Fischer, H.G. 1977, More Ancient Egyptian Names Of Dogs And Other Animals, Metropolitan Museum Journal, Vol. 12, pp. 173–178.

Hanks, P., Coates, R. and McClure, P. eds. 2016, The Oxford Dictionary Of Family Names In Britain And Ireland, Oxford University Press, Oxford, UK.

Hartmann, V.R. 1966, Anwendung der Informationstheorie bei der Schaffung von Markennamen, Jahrbuch der Absatz und Verbrauchsforschung, 4, pp. 326–335.

Hough, C. and Izdebska, D. eds. 2016, The Oxford Handbook Of Names And Naming, Oxford University Press, Oxford, UK.

Keay, A. and Graduand, M. 2011, Emergent Phenomena Of Robot Competitions: robot identity construction and naming, IEEE Workshop On Advanced Robotics And Its Social Impacts, October 2nd–4th, Half Moon Bay, California, USA, pp. 12–15.

Keay, A. 2012, The Naming Of Robots: biomorphism, gender and identity, Master's Thesis In Digital Cultures, University of Sydney, Sydney, Australia.

Krippendorff, K. 2004, Content Analysis: an introduction to its methodology, Sage Publications, Thousand Oaks, California, USA.

Lakoff, G. and Johnson, M. 2008, Metaphors We Live By, University of Chicago Press, Chicago, Illinois, USA.

Mollerup, P. 2013, Wayshowing-> Wayfinding, BIS Publishers, Amsterdam, Netherlands.

Namerobot 2021, Robot Names – naming robots and androids, Retrieved from www.namerobot.com/All-about-naming/All-about-names/Robot-names-androids (Accessed: October 26th 2021)

Newman, L.S., Tan, M., Caldwell, T.L., Duff, K.J. and Winer, E.S. 2018, Name Norms: a guide to casting your next experiment, Personality and Social Psychology Bulletin, Vol. 44, No. 10, pp. 1435–1448.

Parkin, H. ed. 2021, The Concise Oxford Dictionary Of Family Names In Britain, Oxford University Press, Oxford, UK.

Piller, I. 1996, American Automobile Names, Doctoral Dissertation, Technische Universität Dresden, Dresden, Germany.

Rymes, B. 1999, Names, Journal of Linguistic Anthropology, Vol. 9, No. 1/2, pp. 163–166.

Shannon, C.E. 1949, A Mathematical Theory Of Communication, University of Illinois Press, Champaign, Illinois, USA.

Smith, A. 1759, The Theory Of Moral Sentiments, Andrew Millar Publisher, London, UK and Alexander Kincaid Publisher, Edinburgh, UK.

Chapter 6

Meaning

Meaning

Dictionary entries for the word "meaning" usually list at least three concepts:

- the sense or signification of a word or sentence;
- the significance, purpose or underlying truth of something;
- the motive or intention of something.

Dictionary definitions thus suggest that the word "meaning" can be used in two different ways. It can denote something (the first dictionary concept) or it can instead connote attributes and associations of that something (the second and third concepts). A phrase such as "the meaning of the red light is that there is an emergency" would be an example of denoting while a phrase such as "my son's first drawing had great meaning for me" is an example of connoting and associating.

The first way of using the word can be said to be "literal". Feinstein (1982) suggested that "Literal meaning communicates, conveying denotatively...In literal meaning the referents are in one-to-one correspondence. Both the meaning and the referent are consensually agreed upon by the culture, i.e., meaning and referent are already named".

The second way of using the word can instead be said to be "nonliteral". For this manner of usage Feinstein (1982) suggested that "...nonliteral meaning evokes by way of connotation...it primarily evokes and secondarily communicates...Unlike literal meaning in which the referents are in one-to-one correspondence, the referents in nonliteral meaning are in one-to-many correspondence".

Given the two different concepts which can be conveyed, "meaning" can be a mischievous and deceptive little word. It can lead the listener astray. If it is used to suggest what something is referring to, but the listener expects instead attributes and associations, confusion ensues.

DOI: 10.4324/9781003319740-6

Care needs to be taken to embed the word within sentences which constrain to the desired interpretation. The little troublemaker needs to be cornered.

Linguists have tended to focus on the literal sense of the word "meaning". And they have often suggested that meaning can be thought of as a form of belief. Grice (1957) wrote that "perhaps we may sum up what is necessary for A to mean something by X as follows. A must intend to induce by X a belief in an audience, and he must also intend his utterance to be recognized as so intended".

And semioticians have extended that view by analysing the beliefs which are induced not just by words but by any form of representation whether it be natural language, body language, gestures, symbols, mathematics, imagery, clothing, cosmetics, architecture, rituals or other human artefacts or behaviours. Semiotics is in fact the study of signs and symbols as elements of communicative behaviour (Eco 1979; Chandler 2007). In semiotics the word "meaning" usually refers to the denotation or signification produced by the word, symbol or action. The objective is usually to understand the rules and contents of the various forms of human communication.

While perhaps not disputing that meaning can be thought of as a form of belief, sociologists have tended to be less comfortable with the possibility of transmitting it. Social constructivists in particular maintain that reality is not an objective truth but instead a construct of the interaction between people. Social constructivists therefore maintain that language does not mirror reality, but instead creates it. This view (Berger and Luckmann 1966) implies that the transmission of a belief depends on the existence of an a priori linguistic and cultural system, i.e. something can be believed only if it is based mostly on things which were agreed by the people beforehand. On this view, meaning is not so much an objective truth which can be transmitted. Rather, it is more of a jointly constructed and jointly shared subjective belief which arises from the people's interactions.

Libraries are rich in texts which discuss literal meaning. Linguists, sociologists and semioticians share the common interest of wishing to better understand human communication, and human communication relies on the transmission of literal meaning. While not exclusively so, those resources tend to treat the matter of "meaning" from the perspective of communication. While invaluable, literal meaning is not the sense which is most relevant here.

The meaning of "meaning" which is of most interest here is instead closer to English language words such as "purpose", "significance", "motivation", "importance" and "value". It is the nonliteral form which proves more revealing and more rewarding in design. The connoted attributes and associations provide the basis for designing. If not specifically indicated as otherwise, the word "meaning" is used throughout this book in its nonliteral sense.

In its nonliteral sense the word "meaning" can involve the reasons why a person engages with something or someone, what value that engagement might have for the person, and the place that engagement might occupy in that person's subjective realm of life. The connotational use of the word refers to the thought processes and values which emerge from the interaction of memories, emotions and life experiences. It involves the many bits-and-pieces of "purpose", "significance", "motivation", "importance" and "value" which are evoked.

The meanings which people associate specifically with consumer products were analysed by Friedmann and Lessig (1986) who stated that "one can regard consumer behaviour as a continuum ranging from information processing to aesthetics consumption. On the one extreme we can see a logical, methodical information-processor using choice heuristics. At the other extreme we see the consumer aesthetically consuming based upon such feelings as fun, elation, and hedonic pleasure."

Fournier (1991) extended the logic by suggesting that consumer products can be grouped according to the nature of the consumption experience by placing them along the continuum from the utilitarian to the hedonic. He defined eight general categories of consumer meaning: utility, action, appreciation, transition, childhood, ritual enhancement, personal identity and position or role. Adopting a somewhat similar approach, Diller et al. (2005) suggested instead fifteen categories of meaning: accomplishment, beauty, creation, community, duty, enlightenment, freedom, harmony, justice, oneness, redemption, security, truth, validation and wonder.

Such lists permit the observation that they usually comprise nouns rather than verbs, and that the nouns refer not to specific artefacts but instead to unifying concepts. And the various taxonomies of meaning seem to support the idea that meaning is to be found somewhere in the neighbourhood of the "why". In human society the concept of meaning appears to have evolved as a way of grouping together things, ideas and behaviours which are emotionally engaging or considered valuable in some way. The concept appears to be a convenient box which a person, group or whole society can use to store something which they enjoy and value.

Krippendorff and Butter (2007) developed the view that meaning is actually a second-order form of understanding. They suggested that meaning is something which is constructed and that the contexts, activities and interactions within which the construction takes place can best serve to categorise it. They suggested four obvious ways by which artefacts can come to take on meaning for people:

- *a theory of meaning for artefacts in use* which accounts for how people understand and interact with their artefacts in their own terms and for their own reasons;
- *a theory of meaning for artefacts in language* which accounts for how artefacts also occur in conversations among people, not only in user interactions;
- *a theory of meaning for artefacts in their life cycle* which accounts for how an artefact undergoes transformations from its conception to its retirement, and in that process must enrol stakeholders to form networks through which it can travel with ease and direction;
- *a theory of meaning for ecologies of artefacts* which accounts for how different species of artefacts interact with one another, compete or cooperate, and form technological complexes.

Social constructivist views and socially constructing systems are supported by many empirical facts. While not exceptionless general theories, they nevertheless provide useful frameworks for understanding "meaning". And they offer some advantages with respect to those proposals which can be described as being more realist, objectivist or descriptivist in nature.

For example, Csikszentmihalyi and Rochberg-Halton (1981) have shown that meaning can change as a function of the person's age, gender or other demographic descriptors. Studies such as those of Watson (2002) or Wallendorf and Arnould (1988) have instead suggested that the meaning of an artefact can change due to the cultural context in which it is emerged. Much empirical evidence suggests the constructed nature of artefact meaning, and thus that it is often more a matter of the characteristics and culture of the people involved than one of artefact form and function.

While anchored in fact and physics, meanings nevertheless evolve from social interactions and from systems of social construction. They are about the reasons for engaging with something or someone, the value that engagement might have and the place that engagement might occupy in the subjective realm of life. And meanings are of course the meanings of people, and people differ, and change over time. Meanings cannot be considered to be time-invariant universals in the manner of scientific or engineering properties. They are instead beliefs which are continuously emerging and continuously changing.

Meaning In Design

Boradkar (2010) has suggested that "design's core mission is to fashion things so that we may have meaningful interactions with the world. Meanings are neither inherent properties of the things themselves, nor are they total fabrications of the human mind; they are suspended in the spaces between us and all that is around us. Meanings emerge and change continuously as people and things travel through their lives, constantly bumping into each other."

While most artefacts have some meaning for some people, not all of them were designed with meaning in mind. History books are full of tales of things which were developed because, well, because they could be developed. Not all human achievements, perhaps not even the majority of them, are the result of a pursuit of meaning.

However, given the increasing complexity and cost of new products, systems and services it has become more common to consider the meaning and to set targets for its achievement already from the first stages of a design process. The more complex and costly some product, system or service becomes, the more people expect from it. The challenge thus quickly becomes one of finding ways to discuss meaning, to specify it, and to monitor its achievement.

One approach for discussing meaning and for monitoring its achievement was suggested by Giacomin (2017). The core of the framework is a set of three macro-categories of meaning which were identified from literature review. The basic premise is to consider the three forms of meaning, to prioritise one or more of them, and to evaluate their achievement at every point in the design process with at least the major stakeholders. The three macro-categories are: function, ritual and myth.

Function refers to the way something works, to its purpose, or to the duty of some person. Function is a natural way of describing the causalities in the world around us, and is an incredibly natural thought process for human beings. Kelemen (1999) has in fact suggested that "a fundamental aspect of adult thought is the teleological tendency to assume that objects exist for a purpose. When seeing an unfamiliar artefact or strange anatomical part on an animal, the first question an adult will usually ask is what's that for?" Function is perhaps the most commonly used basis for design and is certainly the most frequently reported.

As noted by Maier and Fadel (2009) the concept of "function" is probably so natural to people because it implies action. They also noted, however, that confusion can arise when this is not the case. A nail which holds together a piece of furniture is usually said colloquially to have a function, but while holding things together it does not visibly exhibit an action. Maier and Fadel suggested that "the concept of function also

95

denotes action: the transformation of some input state to an output state. This presents a problem of description for objects which have an obvious use, but no active function transforming inputs to outputs". It is therefore probably worth noting that the everyday use of the word "function" can be positive as in the case of a hammer which acts upon something, or negative as with the nail which avoids the action of the furniture coming apart.

And it is also probably worth noting that an artefact's function does not have to be either simple or unique. Like the Swiss army knife, many artefacts have more than a single obvious function. And some of the functions may have been intentionally designed while others may instead have been unintended uses. Many artefacts end up being used in ways which had not been intended or even imagined by their designers. Artefacts can take on a range of functions depending on the people, places and contexts involved.

To help make sense of multiple functions Achinstein (1983) suggested dividing the possibilities into three categories: design functions, use functions and service functions. He cited the example of a brick which would have been designed to be part of a wall, which could however be regularly used as a door stop, and which might occasionally serve to assist a short individual to reach a doorbell. Achinstein's analysis highlights the fact that something as basic as an artefact's function can prove to be more a matter of who the user is than a matter of its physical characteristics and affordances.

The second macro-category of meaning, that of ritual, refers instead to a series of actions or a type of behaviour which is regularly and invariably performed. It is a set of fixed actions or words, performed consistently and regularly, especially as part of a ceremony. The anthropologist Rappaport (1971) suggested that ritual "is a mode of communication distinguished from other modes of communication by its distinctive codes, namely conventionalised display". And Lukes (1975) defined ritual to be "rule-governed activity of a symbolic character which draws the attention of its participants to objects of thought and feeling which they hold to be of special significance".

Bell (1997) has suggested that the essence of ritual is not the action performed, but, rather, the symbolic meaning it suggests. Six characteristics of ritual acts were proposed: formalism, traditionalism, disciplined invariance, rule-governance, sacral symbolism and performance. And Doty (2000) has drawn attention to the embodied nature of ritual when stating that"the human body itself constitutes an important means of communicating. Its postures, its inborn responses to stimuli, its moods and beauties, its positions in social intercourse: all these may be used in the communicative process and all are utilised in myths and rituals."

A conclusion which can be drawn from such definitions is that they involve adjectives rather than nouns or verbs, and that the adjectives provide constraints to what can be classified as a ritual act. Not just any arbitrary human act can be thought of as being ritualistic, because ritual has requirements. It consists of patterns of action which are emotionally engaging and which communicate in some way. In absence of repetition, exhibition and meaning, an act is not a ritual.

Thinking about an artefact in terms of the ritual or rituals which it may facilitate is common when designing, because rituals are routine parts of everyday life. The morning breakfast, the email session, the lunch break, the evening out, the weekend walk in the park or the weekend religious service are all activities which tend to fall into a ritual pattern, of either a personal or communal nature. And wherever repetitive and emotionally engaging acts are found in everyday life it is usually the case that one or more artefacts are involved. Many designed artefacts act as go-betweens and intermediaries in rituals, connecting the person or persons with themselves, with other people or with other artefacts or environments. Considering the opportunities for facilitating ritual is thus an important and potentially rewarding part of any design activity.

Finally, the third macro-category of meaning, that of myth, refers to a traditional story, especially one about the early history of a people or one which explains a natural or social phenomenon. It is an idealised, exaggerated, and often fictitious conception of a thing or of a person. Mircea Eliade (1963) suggested that myths "provide models of human behaviour and give meaning and value to life". And Levi-Strauss (1955) claimed that myths provide a logical model for overcoming contradictions or paradoxes, that they highlight universal characteristics of our social environment, that they reveal universal human cognitive processes and that they are holistic in nature.

Unlike functions which usually involve some action, or rituals, which are usually about communicating through some action, myth is a category of meaning which consists mostly of a model for action. Myths provide a point of reference and an organising principle. They are webs of connectivities and temporal sequences which when assembled form mental models.

As connected pieces of narrative, myths typically flow from some initial challenge to some final resolution. They transcend the facts, highlighting instead the aspirations, emotions and human significance of the events. They reflect the nature of the human experience, characterised as it is by a continuous immersion in challenges to overcome. Like all models, myths provide a version of events which helps make even the most complicated matters understandable to people through simplicity of narrative.

One of the most complete descriptions of myth was provided by Doty (2000) who suggested seventeen characteristics. Of particular relevance in design are those of being stories, being culturally important, involving emotion, involving participation and conviction, and conveying the political and moral values of the culture. As a category of meaning, that of myth is the one which comes closest to capturing the possible meaning of a work of art, or of a sublime landscape, or of the human values which provide the foundation for ethics. Myths are reference points in design, as in life.

Designing with a specific myth in mind is often a process of seeking to associate the artefact with its source of inspiration. Perceptual stimuli such as colours and sounds from the mythical narrative can serve as design datums, as can any obvious forms or functions. The associations can be sought through similarity of aesthetic or by mimicking some key action from the mythical narrative. Design references to historical narratives, literary stories and cinema are common.

For example, fashion design often involves a declared association with a specific person, movement or style. Complete lines of garments are inspired by, and named after, mythical references. Even when not overtly referencing the mythical inspiration, a choice of colour, or fabric, or shape often provides association with a historical period or a well-known designer. What is the essence of Prada in absence of black?

And in the field of architecture the debates about the style or inspiration of a new building are not idle gossip, but, instead, explorations into how the new construction should be understood by people. With anything as functionally complex and socially relevant as a major building, simplifying the narratives and the points of reference can prove helpful. Emphasising whether the structure should be viewed through a modernist or a post-modernist lens, or noting a foundational moment of the organisation which occupies the building, are ways of making sense of the structure via reference to historical narratives.

And in the automotive sector it is not unusual to seek design inspiration from some well-known figure from automotive history, as talk of the Bentley Brothers or of Enzo Ferrari seems to confirm. Or to design a new automobile starting from the popular and well-understood narrative of an existing vehicle. While facing challenges of translation arising from changes in technology and in societal values, heritage brands in particular make extensive use of historical myths when designing their products and when communicating them to the public. The given auto-mobile may be new, but it is the most recent manifestation in the logical progression from its mythical origin.

Autonomous Road Vehicle Meanings

Given the likely complexity and cost of autonomous road vehicles it will be important to consider the meaning or meanings which the vehicles will occupy in the lives of their users. Clarity of meaning, consistency in achieving that meaning during design, and communication of that meaning to the public will all be vital. The new machines are unlikely to achieve commercial success if confusion is allowed to arise in relation to their connotations, purpose or value.

With human-driven road vehicles more than a century of experimen-tation and of automotive culture has led to several generally recognised meanings. Figure 6.1 provides three relatively pure examples, one cho-sen from each of the three basic categories (function, ritual and myth). It can be noted that the vehicles provide different functions and optimise different characteristics based on their intended role in the lives of the people who use them. While each has a motor, wheels and other similar components, there is nonetheless substantial differentiation in form and function due to the differing human meaning which is supported. The past is, however, the past.

It is therefore interesting to speculate about the future meanings which might be provided. The mechanical nature of 19th and 20th century road vehicles had greatly constrained their design to specific forms and functions. The autonomous road vehicles of the 21st century may instead be freer to offer a wider variety of supports and services. In many cases it may even turn out that mobility itself, i.e. the moving of a person or persons from point A to point B, might no longer be the main element of the meaning of the autonomous road vehicle.

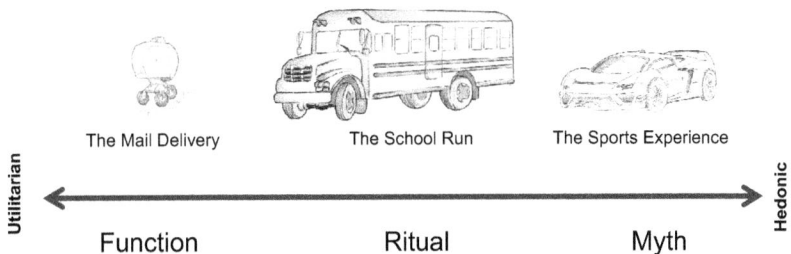

Utilitarian   The Mail Delivery   The School Run   The Sports Experience   Hedonic

Function   Ritual   Myth

Figure 6.1 Examples of traditional road vehicle meanings.

Source: Henry Leeson

By releasing core constraints of 19th and 20th century road vehicles such as the human driving controls, or the access and seating arrangements necessitated by the driver's position, designers will be freer to attempt a wider variety of cabin designs and vehicle packaging arrangements. If electrification were to also free up space and reduce weight, designers will find themselves even further empowered to attempt new vehicle meanings. And if the sensory and computational capabilities of the autonomous road vehicles were to reach levels which permit sophisticated behaviours and social interactions, then the task of designing them will begin to resemble the task of designing a life form.

The elimination of the driving task, the simplifications provided by electrification, the increased computational capabilities and the changes in societal attitudes may all be converging towards a disruptive shift in the concept of road vehicle. It may transcend its past roots as a mechanical machine for providing mainly point-to-point mobility, becoming instead more of a "friendly neighbourhood robot" which can assist any number of functions, rituals or myths.

Consideration of the meanings which autonomous road vehicles might support suggests the likelihood of a spectrum of machines whose most obvious meaning ranges from the very functional to the very mythical, as illustrated in Figure 6.2. Along the spectrum there are likely to be some autonomous road vehicles which share some similarities in meaning to some existing road vehicles. But changed or completely new meanings are also likely, and are not difficult to find on the pages of current newspapers and magazines. It has not escaped the public's attention that improvements in automation and in artificial intelligence are transforming road vehicles from the mechanical to the social, i.e. from

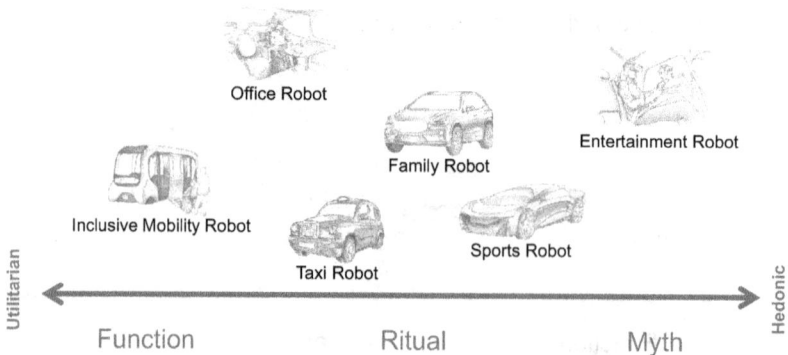

Figure 6.2 Examples of autonomous road vehicle meanings.

Source: Henry Leeson

machines to partners. Whereas traditional road vehicles could only implement driver or passenger requests, there is a growing public awareness
that autonomous road vehicles also interpret, and sometimes anticipate,
those requests. And with the growing awareness of the autonomous
road vehicle's social dimension come speculations as to what might be
provided and what might be accomplished by such a social partner.

One non-traditional functional meaning which is a topic of current discussion is that of "battery robot". It has been repeatedly suggested that
the batteries of future electrified autonomous road vehicles can provide
electrical storage backup for electrical power grids, feeding electricity
back at night when not needed by the vehicle, or perhaps even powering
appliances or complete homes for short periods during the day in cases
of need. While possibly providing some human or goods transport, such
robots would be carefully optimised in terms of their electrical power
storage and grid connectivity. A "battery robot" was not easily achieved
using the technologies of the 20th century but experts suggest that it
will be part of our 21st century future. Some of the upcoming friendly
neighbourhood robots are likely to be less about transport and more
about electricity supply and emergency response. The meaning of such
friendly neighbourhood robots will be more about energy than about
transport.

Another emerging meaning is that of "inclusive mobility robot". Such
a meaning is not entirely new with road vehicles, but takes on greater
significance with friendly neighbourhood robots. Language is pregnant
with phrases which refer to friction, barriers and exclusion. Metaphors
such as "being born on the wrong side of the tracks" attempt to convey
the experiences of people who encounter impediments to their rights or
ambitions. And terms such as oligarchism, elitism, racism, sexism, col-
ourism and ableism illustrate the variety of characteristics and conditions
which can be used to erect obstacles.

Autonomous road vehicles which are designed specifically to optimise accessibility, inclusivity and human dignity seem a reasonably certain prospect given their likely flexibility and possible cost effectiveness.
Where the mechanicals and driver's position placed severe constraints
on past designs, the greater packaging freedom and computational
resources of friendly neighbourhood robots should make it possible to
accommodate individuals who declare a range of needs or desires. From
"special needs robots" to "multi-user low cost robots" there is currently
much speculation about the level of accessibility and inclusion which
can be provided by unconstrained autonomous road vehicle cabins, and
by the cost reductions which may occur due to the absence of expensive certified human drivers. In all these cases the meaning, while not

completely new, shifts greatly in the direction of barrier reduction and personal freedom.

And since improvements in automation and in artificial intelligence are transforming road vehicles from the mechanical to the social, i.e. from machines to partners, there appear to be opportunities for social interaction and perhaps even social care. It does not seem unreasonable to suggest that some future forms of "inclusive mobility robot" will support not just the physical transport needs but also some of the social needs of their passengers. Perhaps even from within a healthcare or social care setting. And perhaps the supports and freedoms afforded by such friendly neighbourhood robots will stimulate emotions and enthusiasms which are reminiscent of those of the early 20th century users of road vehicles, for whom the new form of transport shortened distances and expanded the personal and professional horizons.

Finally, one new meaning which is also a subject of much current speculation is that which results from the blending of transportation and entertainment. An "entertainment robot", referred to by some as "transportainment" (Beyer 2016; Schukert and Müller 2006), would involve providing specialised entertainment functions in addition to providing the traditional point-to-point mobility. While a variety of minibuses, tourmobiles and other road vehicles are currently used for purposes of entertainment, leisure or tourism, their meanings are mainly anchored around their transport function. The didactic or entertainment elements of the experiences are usually currently provided by human professionals, not by the vehicle itself.

Future "entertainment robots" may instead be autonomous also in the non-driving aspects of the experience. The capable computerised companions may perform the multiple tasks of a single human, and perhaps even the multiple tasks of multiple humans. Since improvements in automation and in artificial intelligence are transforming road vehicles from the mechanical to the social, multiple human interactions will be possible which have little or nothing to do with physical transport. Gaming robots, party robots, tourism robots, well-being robots and other such propositions are recurring themes. And as examples such as the Star Trek holodeck or the Westworld robots suggest, catering for human entertainment desires has been a constant feature of science fiction literature over the years (Shedroff and Noessel 2012) thus demonstrating the likely demand for such services. Given the many concepts which have already been socially explored by science fiction it would be surprising if some of them did not become manifest in science fact.

Imagining what might be possible when the road vehicle becomes less of a machine and more of a partner is an interesting intellectual endeavour. And with time, business opportunities will congeal into societally recognised functions, rituals or myths. And much of the responsibility for the ideation, articulation and concrete manifestation of the new meanings will fall to the designers.

As the first friendly neighbourhood robots enter service there will be as great a need to make them meaningful as there will be to make them work safely and efficiently. Public acceptance of a machine as functionally complex and socially relevant as a friendly neighbourhood robot will depend greatly on the clarity of its meaning and on its role in people's lives. Thus the early work of designers may end up being as much about "meaningfication" as about optimising the functional properties of the vehicle.

When a designer identifies an opportunity which interconnects several previously unrelated technological and societal trends, and articulates one or more product, system or service concepts which address the opportunity, the process can be called "meaningfication" (Giacomin 2017). The process involves data, ethnography, real fictions and co-creation to interconnect patterns of technology, experience, personal identity, societal identity, value assignation and consumption. In a world which is ever richer in products, systems and services, the process of "meaningfication" takes on great importance as new commercial spaces are sought. Commercial opportunities are about people, and people are highly sensitive to meaning.

Given the early stage in the history of the friendly neighbourhood robots, meaningfication may prove to be one of the biggest current hurdles to overcome. Design work is thus likely to involve a large number of activities which seek out the possible new meanings and which test the ability of the proposed friendly neighbourhood robots to provide them. Given the early days of the relationship, much work is still needed to understand what humans want from their friendly neighbourhood robots and to understand if it can be provided.

Conclusions

Meaning can refer to the significance, purpose or underlying truth of something. Such meaning is to be found somewhere in the neighbourhood of the "why". It is about the reasons why a person engages with something or someone, what value that engagement might have, and the place that engagement might occupy in the subjective realm of life. This chapter has provided a short introduction to the concept of meaning, to meaning in design and to autonomous vehicle meanings.

It was noted that the meanings which people associate with consumer products have been investigated by several researchers and that they have usually been found to lie along a continuum from the utilitarian to the hedonic. Information processing behaviours on the part of the consumers have been suggested to anchor one extreme of the continuum, while aesthetics consumption has been suggested to anchor the other.

In this chapter several taxonomies were introduced which group the possible forms of consumer product meaning into categories based on their general nature. One particularly complete example, first suggested by Diller et al. (2005), consists of fifteen categories of meaning: accomplishment, beauty, creation, community, duty, enlightenment, freedom, harmony, justice, oneness, redemption, security, truth, validation and wonder.

And beyond the general nature of the consumer product meanings it was suggested that it can also prove helpful to consider the systems within which those meanings emerge. As forms of constructed belief, the contexts, activities and interactions within which the meanings emerge can serve to categorise them. And Krippendorff and Butter have suggested four obvious ways by which artefacts come to take on meaning for people: artefacts in use, artefacts in language, artefacts in their life cycle and ecologies of artefacts.

This chapter introduced a design framework for discussing meaning and for monitoring its achievement during the design process. The framework, first suggested by Giacomin, is based on three macro-categories of meaning: function, ritual and myth. The properties of each of the three macro-categories were described and examples of each were provided. It was claimed that it is possible to increase the focus and consistency of the design process by first identifying the exact form, or exact mixture, of meanings which are intended for an artefact.

This chapter repeatedly noted that releasing core constraints of 19th and 20th century road vehicles such as the human driving controls, or the access and seating arrangements necessitated by the driver's position, will allow designers to attempt a wider variety of cabin designs and vehicle packaging arrangements. And that, with time, these new opportunities would permit new road vehicle meanings.

From among the concepts which are currently being debated in relation to autonomous road vehicles, three were selected as illustrative examples of the possible new meanings which road vehicles may assume in the future. Each is a stereotypical example of one of the three basic categories of meaning (function, ritual and myth). And each was described in terms of its basic logic and usage. The three concepts were the "battery robot", the "inclusive mobility robot" and the "entertainment robot".

Finally, it was noted that when the first friendly neighbourhood robots enter service there will be as great a need to make them meaningful as there will be to make them work safely and efficiently. Societal acceptance will depend on the clarity of the friendly neighbourhood robot's meaning and on its role in people's lives. The concept of "meaningfication" was introduced to describe the design activities required to clarify meaning, and to design for it. And it was suggested that current autonomous road vehicle design work is likely to be involving a large number of such activities which seek out the possible new meanings and which test the ability of the proposed friendly neighbourhood robots to provide them.

Having introduced the topic of meaning and its importance for autonomous road vehicles, the next chapter introduces a design tool required for its achievement: metaphor.

References

Achinstein, P. 1983, The Nature Of Explanation, Oxford University Press, Oxford, UK.

Bell, C. 1997, Ritual: perspectives and dimensions, Oxford University Press, Oxford, UK.

Berger, P.L. and Luckmann, T. 1966, The Social Construction Of Reality: a treatise in the sociology of knowledge, Anchor Books, Garden City, New York, USA.

Beyer, A. 2016, Le transport fait-il partie du voyage? Pour une compréhension du déplacement touristique à partir de l'antagonisme contrainte/agrément, Géotransports, Vol. 7, pp. 7–22.

Boradkar, P. 2010, Designing Things: a critical introduction to the culture of objects, Bloomsbury Academic, London, UK.

Chandler, D. 2007, Semiotics: the basics, Routledge, Abingdon, Oxfordshire, UK.

Csikszentmihalyi, M. and Rochberg-Halton, M. 1981, The Meaning Of Things, Cambridge University Press, Boston, Massachusetts, USA.

Diller, S., Shedroff, N. and Rhea, D. 2005, Making Meaning: how successful businesses deliver meaningful customer experiences, New Riders Publishing, Berkeley, California, USA.

Doty, W.G. 2000, Mythography: the study of myths and rituals, University of Alabama Press, Tuscaloosa, Alabama, USA.

Eco, U. 1979, A Theory Of Semiotics, Indiana University Press, Bloomington, Indiana, USA.

Eliade, M. 1963, Myth And Reality, translated by W. Trask, Harper and Row, New York, New York, USA.

Feinstein, H. 1982, Meaning And Visual Metaphor, Studies in Art Education, National Art Education Association, Vol. 23, No. 2, pp. 45–55.

Fournier, S. 1991, Meaning-Based Framework For The Study Of Consumer–Object Relations, Advances in Consumer Research, Vol. 18, pp. 736–742.

Friedmann, R. and Lessig, V.P. 1986, A Framework Of Psychological Meaning Of Products, in Lutz, R.J. and Provo, U.T. (Eds.), 1986, North American Advances in Consumer Research, Volume 13, Association for Consumer Research, pp. 338–342.

Giacomin, J. 2017, What Is Design For Meaning?, Journal Of Design, Business & Society, Vol. 3, No. 2, pp. 167–190.

Grice, H.P. 1957, Meaning, The Philosophical Review, Vol. 66, No. 3, July, pp. 377–388.

Kelemen, D. 1999, Function, Goals And Intention: children's teleological reasoning about objects, Trends In Cognitive Sciences, Vol. 3, No. 12, pp. 461–468.

Krippendorff, K. and Butter, R. 2007, Semantics: meanings and contexts of artifacts, In Schifferstein, H.N.J. and Hekkert, P. (Eds.), Product Experience (pp. 353–376), Elsevier, Amsterdam, The Netherlands.

Lévi-Strauss, C. 1955, The Structural Study Of Myth, The Journal Of American Folklore, Vol. 68, No. 270, pp. 428–444.

Lukes, S. 1975, Political Ritual And Social Integration, Sociology, Vol. 9, No. 2, pp. 289–308.

Maier, J.R. and Fadel, G.M. 2009, Affordance Based Design: a relational theory for design, Research In Engineering Design, Vol. 20, No. 1, pp. 13–27.

Rappaport, R.A. 1971, Ritual, Sanctity, and Cybernetics, American Anthropologist, New Series, Vol. 73, No. 1 pp. 59–76.

Schukert, M. and Müller, S. 2006, Erlebnisorientierung im touristischen transport am beispiel des personenluftverkehrs. In Weiermair, K. and Brunner-Sperdin, A., Erlebnisinszenierung Im Tourismus (pp. 153–166), Erich Schmidt Verlag, Berlin, Germany, pp. 153–166.

Shedroff, N. and Noessel, C. 2012, Make It So: interaction design lessons from science fiction, Rosenfeld Media, Brooklyn, New York, New York, USA.

Wallendorf, M. and Arnould, E.J. 1988, My Favorite Things: a cross-cultural inquiry into object attachment, possessiveness, and social linkage, Journal Of Consumer Research, Vol. 14, No. 4, pp. 531–547.

Watson, J., Lysonski, S., Gillan, T. and Raymore, L. 2002, Cultural Values And Important Possessions, Journal Of Business Research, Vol. 55, pp. 923–931.

Chapter 7

Metaphor

Metaphor

Dictionary entries for the word "metaphor" usually list at least three concepts:

- an expression often found in literature that describes a person or object by referring to something that is considered to have similar characteristics to that person or object;
- a figure of speech in which a word or phrase is applied to an object or action that it does not literally denote, in order to imply a resemblance;
- a thing regarded as representative or symbolic of something else.

Metaphor is thus the understanding of one thing in terms of another. It is a way of comparing between objects, people or ideas, with the comparison usually being between something simple or familiar on the one hand and something more complex or unfamiliar on the other. The referencing to the simpler suggests commonalities and structures, facilitating the understanding of the more complex. Most human thought processes are said to be metaphorical and the metaphors are not arbitrary, they are instead based on physical and cultural experience (Lakoff and Johnson 1980).

Probably the most often cited description of metaphor in language is that of Black (1955) which consists of seven requirements:

- a metaphorical statement has two distinct subjects: a "principal" subject and a "subsidiary" one;
- these subjects are often best regarded as "systems of things" rather than "things";
- the metaphor works by applying to the principal subject a system of "associated implications" characteristic of the subsidiary subject;

DOI: 10.4324/9781003319740-7

- these implications usually consist of "commonplaces" about the subsidiary subject, but may, in suitable cases, consist of deviant implications established ad hoc by the writer;
- the metaphor selects, emphasises, suppresses, and organises features of the principal subject by implying statements about it that normally apply to the subsidiary subject;
- this involves shifts in meaning of words belonging to the same family or system as the metaphorical expression, and some of these shifts, though not all, may be metaphorical transfers;
- there is, in general, no simple "ground" for the necessary shifts of meaning – no blanket reason why some metaphors work and others fail.

And any statements which meet the criteria to be considered metaphors can then be further analysed by subdividing them according to the comparison which is made. According to Lakoff and Johnson (1980) any linguistic metaphor can be classified into one of three general categories: ontological, orientational and structural.

An ontological metaphor is one in which something abstract such as an idea or an emotion is compared to something more concrete such as a substance, object or person. Ontological metaphors are attempts at providing familiar real-world characteristics to concepts which are usually not physical, thus not directly perceivable through sensory experience. Often, it is a way of suggesting shape and substance to something which has neither. Phrases such as "he is in love" or "necessity is the mother of invention" are examples.

An orientational metaphor involves instead a spatial arrangement such as up-and-down, front-and-back or in-and-out. Orientational metaphors assign direction and allude to priority, thus they serve to organise thinking. As Lakoff and Johnson noted, many concepts such as good, bad, happy, sad, positive and negative are orientated. For example, in most languages the word happy is associated with the upward direction while the word sad is associated with the downward direction, as in the phrase "I am in high spirits" or "I fell into depression".

Finally, a structural metaphor is one in which a complex concept is described in terms of a simpler concept in order to highlight key characteristics. Structural metaphors act as focussing tools. They usually suggest familiar properties and principles from the simpler concept which can help us to better understand the more complex concept. A phrase such as "argument is war" is an example in which it is suggested that the concept of "argument" can be thought about or analysed in a similar manner to that of a "war".

Understanding what can be legitimately considered a metaphor is helpful, but does not fully explain how people make sense of the comparisons. How a linguistic metaphor actually works in practice is even today something of an open question. Nevertheless, a few attempts have been made to describe the possible logical comparisons which are made.

One influential analysis of linguistic metaphors was performed by Turner (1987) who suggested ten inference patterns to explain how aspects of a metaphor source might be associated by people to the metaphor target. He referred to the inference patterns as kinship relations, which account for what the metaphor source and target have in common. The ten suggested inference patterns were: property transfer, similarity, group, inheritance, components and contents, order and succession, causation as progeneration, biological resource as parent, place and time as parent, and lineage in the world the mind and behaviour. Each pattern accounts for a bundle of properties which can be connected from the source to the target in people's minds. Each is a specialised filter or viewing aperture through which things may appear similar. Thus each indicates possible ways of selecting metaphors and of applying them.

And metaphors are not limited to those which are linguistic. Metaphors can be used in any form of semiotic communication whether it be visual, acoustic, haptic, olfactory or gustative in nature. Regardless of the sensory modality through which it acts, all metaphors involve the understanding of one thing in terms of another. And the referencing of the simpler is usually used to suggest commonalities and structures which facilitate the understanding of the more complex.

Visual metaphors, also known as pictorial analogies, are particularly common in everyday life due to their extensive use in advertising. A mental linking is stimulated between the metaphor source image and the target product, system or service. The choice of the metaphor source image is usually made based on its ability to visually express the key physical or conceptual properties which the advertisement intends to capture. The selected source image provides a web of denotations and connotations which help to convey the key messages of the advertisement.

When two images are used simultaneously in an advertisement there are various options for the visual composition, each with its own strengths and weaknesses in terms of conveying the advertising message. Phillips and McQuarrie (2004) have suggested that the images can be arranged in three possible ways: "juxtaposition" where the source and the target are depicted alongside each other, "fusion" where the source and the target are merged together visually and "replacement" where the target is completely replaced by the source. And Phillips and

Figure 7.1 A visual metaphor used in automotive advertising.

Source: Concept by Africa; Image by Platinum FMD

McQuarrie (2004) have also suggested that the underlying logic of the visual composition can be one of three types: connection ("A is associated with B"), similarity ("A is like B") or opposition ("A is not like B").

Figure 7.1 is an example of the use of visual metaphor in automotive advertising which involves juxtaposition and similarity. The joint presence of both the metaphor source and the metaphor target stimulate a particularly immediate and direct understanding of one in terms of the other. The animal was presumably selected to be representative or symbolic of the road vehicle. Thus people's familiarity with the animal is exploited to imply key characteristics such as power, strength and durability for the road vehicle.

Metaphor In Design

Design involves the deciding of many physical, perceptual, cognitive and emotional characteristics of the product, system or service. The number of relevant parameters is potentially unlimited. Tools which can help to simplify, select and communicate are thus extremely helpful. And metaphor is among the most important of those tools.

The adoption of a central coordinating metaphor provides a focal point and reference for design. It helps to clarify what makes sense to consider and what instead does not belong. Characteristics which do not fit the central metaphor of the design are likely to lead to misunderstandings on the part of the users, thus are best avoided. Characteristics which instead support the central metaphor of the design are likely to increase the safety, efficiency, interaction and emotional engagement. A central coordinating metaphor helps to keep things simple and keep them focussed.

The importance of metaphor in design was noted by Casakin (2007), who suggested that "design is a complex and ill-structured activity, where problems cannot be solved through the application of algorithms or operators". In Casakin's view, metaphors guide reasoning and assist designers by helping to capture the most important concepts, goals and requirements. While not scientifically rigorous or universally valid, a familiar metaphor can provide a powerful tool for coordinating the many design decisions which need to be taken.

Torgny (1997) suggested that "creative processes often involve a combination of familiar concepts into new ones and here metaphors can play an important role. The use of common metaphors, for example in a working group, can help to intensify the dynamics within the group but also to stimulate new visions and concepts."

And Schön (1993) proposed the concept of "generative metaphor" which he claimed referred to both "a certain kind of product – a perspective or frame, a way of looking at things – and to a certain kind of process – a process by which new perspectives on the world come into existence".

Writing specifically about product design, Cila et al. (2014) suggested two properties of the metaphor source which should be considered when searching for one to use: salience and relatedness. Selecting a highly salient and obviously related metaphor source can render the design process simpler, easier and more communicable by focussing attention on key characteristics and capabilities.

In their view, the term "salience" refers to the purity of the structure which the source offers. Choosing an automotive tyre as the metaphor source for an automotive steering interface can be considered an example of salience. Choosing a hand-operated wheel for steering the vehicle's tyres offers a shape (circular) and motion (rotational) which are similar, thus the relationship would be expected to be obvious to most people. Confusion in relation to the control action is unlikely.

Also, in their view, the term "relatedness" refers to the overall conceptual closeness between the chosen metaphor source and the target artefact. For example, a metaphor source such as an automobile would be close to a target artefact of a van, but relatively far from a target artefact of a chair or a lamp. If the metaphor source and the metaphor target are close, people are unlikely to get confused about the general category of the artefact, its intended purpose or its manner of use.

The use of a well-chosen metaphor as an organising tool is particularly helpful when designing interfaces. With any product, system or service the user interface is the point of contact between the artefact and the human, and thus greatly influences the perceptions and interactions which occur. Public acceptance and commercial success often depend critically on the intuitiveness and ease-of-use of the artefact's interface. And a well-chosen metaphor helps to achieve intuitiveness and ease-of-use.

The role of metaphor in the design of interfaces was investigated by Celentano and Dubois (2014) who defined three measures of transfer from a source concept to a target interface: coherence, coverage and compliance. Each describes a way in which the source concept can help the user to learn about the interface and to use it.

Coherence was defined as the degree of transfer from the source concept to the target interface of the physical structure and functional capabilities. The term coherence is used to capture the idea that an interface will prove more intuitive and easier to use if its characteristics and functions closely match those of the metaphor source. If the thing or place or behaviour which serves as the source for the metaphor has certain elements, then the interface would benefit from having as many of them as possible and from not having elements from other things, places or behaviours. Coherence is about people finding the kinds of things they expect based on their understanding of the underlying metaphor. If a designer chooses a taxi driver (i.e. the human person) as the metaphor source for a vehicle interface then users might become confused if the interface were to offer functions for purchasing groceries or for vacuum cleaning the house.

Coverage was instead defined as the total number of characteristics from the source concept which are achieved in the target interface. The term coverage is used to capture the idea that an interface will prove more intuitive and easier to use if its characteristics and functions match as many as possible of those which are present in the metaphor source. If for example a designer were to choose a taxi driver as the metaphor source for a given vehicle interface then users might become confused if there were no interface option for indicating where to be dropped off at journey's end. Confusion would arise if the interface seemed generally consistent with a taxi service, but was instead providing only fixed stops in the manner of a bus, tram or train.

Finally, the third measure, that of compliance, was defined as the degree to which the affordances which were present in the source concept are perceived by people in the target interface. The term compliance is used to capture the idea that an interface will prove more intuitive and easier to use if people can understand what it can do, and how to do it, in much the same manner as they would have understood those things with the metaphor source. If for example a designer were to choose a taxi driver as the metaphor source for a vehicle interface then users might become confused if hand clapping was required to request a drop off rather than asking for it verbally as with a human taxi driver.

For computer interfaces specifically, Carroll and Mack (1999) have claimed that a well-chosen and well-communicated metaphor stimulates the user's self-generation of a mental model of the interface. The design metaphor guides the user towards a mental model by drawing attention to critical elements of the interface (see Figure 7.2 below). It provides characteristics and structures which can be used to analyse the interface and to monitor the outcomes which result from its use. Beyond its role as a design tool, metaphor can thus also be considered a learning tool which supports the formation of memories and the emergence of mental models.

It can be stated with confidence that metaphors are widely used in design and that an obvious metaphor greatly simplifies the design process. Particularly during the concept ideation phase of a project, the use of a metaphor source narrows down the options to a much smaller number of parameters and characteristics from which to choose. As a focussing lens, metaphor helps to narrow the infinite down to the finite, providing structure and constraints along the way. Metaphor is a natural basis for human thinking. And metaphor is a natural basis for design.

Autonomous Road Vehicle Metaphors

Several well-known metaphors emerged for road vehicles over the course of the 20th century. Figure 7.3 illustrates a few of the more

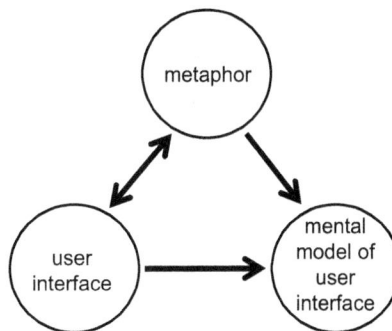

Figure 7.2 Metaphor as a guide to the mental model of a computer interface.

Source: adapted from Carroll and Mack 1999

Company Car

Rental Car

Taxi

Family Car

Figure 7.3 A few traditional 20th century automobile metaphors.

Source: Henry Leeson

traditional overall metaphors which the reader is likely to be familiar with for automobiles. It is noteworthy that the automobiles themselves often differ relatively little in terms of physical or functional characteristics, thus a change in usage pattern or social significance can be the main cause of the change in metaphor.

As suggested by the examples of Figure 7.3 the physical characteristics and functions of the road vehicle provide only some of the elements of its metaphor. Metaphor is a word used to describe a human mental capacity which serves as a basis for thinking, thus it does not exist outside the minds of people and cannot be a property of the artefact alone. Metaphors emerge from interactions with things or people within contexts, and become embedded within thought and memory. As the well-known automobile metaphors of Figure 7.3 suggest, the usage context and the people involved are often more critical to the metaphor and to its significance than the technical characteristics of the vehicle. For example, the main difference between a company car metaphor and a rental car metaphor may simply be the person who is paying the bill.

As Black (1955) noted, metaphors are systems of characteristics, ideas and implications. A metaphor is a bundle of thoughts which is not necessarily fully open to scrutiny and scientific explanation. The colour green can be easily understood to be part of a metaphor for environmentalism given the scientific fact that much of the earth's vegetation is green. Much less obvious, instead, is the fact that a diamond ring should be a sign of marriage and often part of its metaphor. While many metaphors are the direct result of physical facts, others are more subtle allusions to the language, culture or values which a person or a society has evolved over time.

At the time of writing of this book it is not yet possible to discuss autonomous road vehicle metaphors with confidence. The reason is simple, they are not yet a commonly encountered commercial reality. Only informed speculations can be made about autonomous road vehicles until such time as a lot more of them are around. Metaphors develop from people, interactions and time.

At present it is however possible to anticipate that the elimination of the driving task and thus of the driving position, the simplifications provided by electrification and the changes in societal attitudes will probably lead to a shakeup in the set of metaphors which we use to think about road vehicles. The electrified, computerised and versatile friendly neighbourhood robots of the 21st century should be able to provide a wider variety of supports and services, leading to a wider variety of metaphors.

At present it is also possible to note that the improvements in the algorithms which monitor, interpret and make decisions are leading to road vehicles which can be categorised more in terms of the service which they provide than in terms of their on-road or off-road capabilities. Where some 20th century vehicle metaphors emerged from obvious differences in physical size, cabin layout, maximum speed or road holding ability, it seems instead likely that the service provision will provide a more obvious differentiator of 21st century autonomous road vehicles. Going forward, it may be the case that the overall metaphor of the vehicle will be more about what it is trying to provide than about where it goes or how it drives.

And it is possible at the present time to direct attention to a few of the currently debated speculations (see for example Wayner 2013 for a review of some current proposals) so as to note things which may be part of the future metaphors. Much has already been written about the possibilities, often based on a logic of "the technology will be capable of doing this". While not all current speculations appear technically, economically or environmentally sustainable, some are not distant from existing products, systems or services which have already proven successful. Given the similarities in meanings and metaphors to existing designs, such opportunities would seem likely to obtain a degree of success going forward. The discussion below is intentionally limited to a small number of examples chosen from the more likely and less disruptive possibilities.

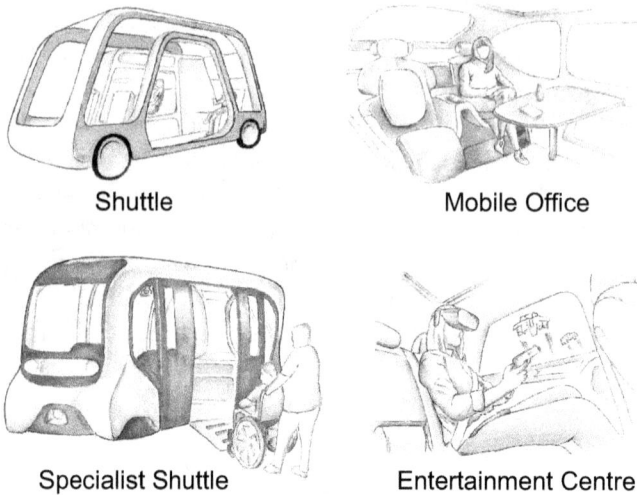

Shuttle   Mobile Office

Specialist Shuttle   Entertainment Centre

Figure 7.4 Possible future autonomous road vehicle metaphors.

Source: Henry Leeson

Figure 7.4 presents four autonomous road vehicle metaphors which are currently much discussed. They are overall metaphors, i.e. holistic understandings of the complete vehicle in terms of its nature, abilities and role in people's lives. Each is a general-level understanding of the vehicle, each is grounded in well-known human needs and each is likely to be socially accepted in the future.

An autonomous shuttle or autonomous pod is not distant from existing airport shuttle busses or trams in terms of function, ritual experience or mythical meaning associated with the travel. There is an air of "incremental" to such offerings. Autonomous road vehicles of this type are likely to be among the first of our friendly neighbourhood robots to reach mass production and to begin supporting humanity. Short and clearly defined routes simplify the constraints and reduce the technical self-driving challenges. And carrying capacity and ease-of-access clearly communicate the purpose of transporting a range of humans of various size, shape and other characteristics. The target metaphor of such offerings does not seem distant from that of current shuttle busses and trams, and any new characteristics would be expected to be easily accommodated via extension of the existing metaphors.

Developing an autonomous office is also not a large leap of imagination from some of the better equipped shuttle busses and train services of today. Again, there is an air of "incremental" to such offerings since such an autonomous road vehicle can be seen as a technological response to spending more time working outside of the traditional office. If the ratio of "time spent travelling to activities" to "time spent at a fixed location" were to increase substantially, then such autonomous road vehicles would prove beneficial as a way of optimising work efficiency and reducing costs. The target metaphor of such offerings would probably be a somewhat more complex version of that of the shuttle, involving similar human transport characteristics, but extending the transport logic by incorporating further characteristics and functions from office environments. Some merging of the current metaphors of "office" and "shuttle" may prove to be a logical starting point in the development of such friendly neighbourhood robots, with new characteristics and functions being added later as societal needs and attitudes evolve.

Autonomous transport for use by specialist communities such as hospitals, care homes or retirement villages is also a popular current proposal. Increasing lifespan, increasing quality of life expectations, increasing periods of illness, increasing medical costs and other factors are leading to increasing requests for transport by individuals with specialist needs. In this case a larger leap of imagination is required to establish what the optimal target metaphors might involve. The key issue is that of "specialist need", which of course depends on the exact demographic, medical profile and social profile of the person or persons involved. The needs and desires of the people who constitute the specialist community will dictate the product specifications of these vehicles, leading to unique requirements which will have to be prioritised with respect to the traditional transport capabilities. Given the increasingly blurred boundaries between healthcare, social care, lifestyle and leisure, there may prove to be a great variety of current metaphors which, when mixed in appropriate proportions, might meet the needs of a given group or specific community. Of the four autonomous road vehicle metaphors which were selected for discussion here, the friendly neighbourhood robot in support of specialist communities is likely to be the one which exhibits the greatest variation in design and commercialisation, going forward.

Of the four complete vehicle metaphors which were selected for discussion here, the least familiar is perhaps that of entertainment centre. At present there are many human-driven vehicles which are designed for purposes of entertainment, leisure or tourism. From sightseeing busses to Nashville street parties (Rojas 2021), road vehicles are used for what has recently been described as "transportainment" (Beyer 2016; Schukert and Müller 2006), i.e. the blending of transport and entertainment. Suggestions have been made in recent years that autonomous road vehicles will provide additional opportunities for road vehicle based entertainment and leisure. The elimination of a costly and dedicated human driver is suggested to lower the expense, and increase the privacy, of a range of entertainment services. The premise is that an affordable mobile party can act as a multiplier of the enjoyment derived from similar activities at a fixed location, the motion and mobility providing additional sensory and cognitive dimensions to the experience. Whether the multiplying effect will be confirmed in practice is as yet unknown, however, it is not difficult to imagine a target metaphor of a 1980s video arcade on wheels or a small exclusive nightclub for use as part of a night out on the town. Despite the less familiar nature, also in this case there seem to be some precedents for the new services which autonomous road vehicles may provide.

If the friendly neighbourhood robots are to serve humans, then their first generation will likely perform tasks which people are already asking for. The overall vehicle metaphors will likely leverage familiar ones which we grew up with and thus feel comfortable with, helping us to transition from the human-driven past to the friendly robot future. In many cases the overall metaphor of the friendly neighbourhood robot may be the same as that of an existing human-driven road vehicle, but stretched and expanded to accommodate new capabilities. The same general idea, only better, safer and perhaps cheaper.

In fact, services such as traditional taxi fleets have been repeatedly proposed to be among the early adopters of friendly neighbourhood robots. While challenging technically, the handing over of the human driver's responsibilities to the automation is straightforward psychologically. Such a transition should not prove particularly confusing to customers and is not likely to significantly disrupt the traditional taxi metaphor. Autonomous shuttle services have also been proposed for the purpose of linking rail or air hubs with other forms of transport at car parks, bus depots, micro-mobility clusters or walking paths. Improving safety and cost, while plugging gaps in travel networks, would appear to be low-risk early roles for the friendly neighbourhood robots. Rather than be replaced, some current overall vehicle metaphors will be reworked.

And appropriate design metaphors are likely to prove important not just for the complete vehicle but also for individual components and subsystems. An example of the importance of subsystem metaphors is the case of the automotive dashboard. Twentieth century designs integrated a small number of objects which usually included the speedometer, odometer, fuel gauge, engine temperature gauge, radio and heating controls. However, 21st century designs additionally include situation awareness displays, infotainment system screens and sensors to perceive things such as driver eye angles, hand movements and body postures. Automotive dashboards have become increasingly more difficult and time consuming to design as the additional systems have competed for the available space. Experiments are thus underway to group and simplify the arrangements using new metaphors which establish new roles and responsibilities for that area of the cabin. Dashboard design appears to be moving away from traditional metaphors such as that of the "instrument cluster" to more advanced metaphors such as that of "situation awareness system", i.e. from metaphors based on the engineering monitoring of the vehicle's mechanical systems to metaphors based more on the understanding of multiple relevant aspects of the situation within and outside of the vehicle.

Design metaphors for autonomous road vehicle components and subsystems are in fact urgently needed. Human-driven automobiles interacted with people mainly through their primary controls and dashboard, but future autonomous road vehicles are likely to have fundamentally different subsystems. Human requests and opinions will be solicited in new ways via new systems which will not initially benefit from metaphors which were established through past commercial and societal practice. New bundles of thoughts are needed to facilitate design and use.

For example, the autonomous road vehicle's Internal HMI System and External HMI System (see Chapter 2 for details) are each likely to have a distinct interface which acts as a fixed point of reference for humans. Each of these interfaces is likely to be new to people, and thus to not benefit from an existing automotive metaphor in support of thought processes. The design of each will thus require a metaphor, either borrowed from a non-automotive environment or established ex novo, which can help to simply the design decisions and coordinate the characteristics. Well-chosen and well-communicated metaphors for the Internal HMI and External HMI Systems will prove decisive towards achieving safety, efficiency, interaction and emotional engagement.

The importance of appropriate metaphors can be appreciated by considering the passenger interactions needed for route planning and route following. Whether it be a computer screen, a loudspeaker or some form of robotic face or body, a reference point is needed inside the vehicle to clarify where the passengers should be directing their vocal or gestural instructions and where they might expect the vehicle's requests and communications to come from. A single coordinating metaphor will help to ensure the consistency of location, physical shape, colour, visual emissions, acoustic emissions and other design parameters of the reference point.

If a face on a screen were chosen for the route planning and route following point of reference, why would passengers look or direct their speech elsewhere when giving directions or when seeking assistance?

Conclusions

Metaphor is a way of comparing between objects, people or ideas, with the comparison usually being between something simple or familiar on the one hand and something more complex or unfamiliar on the other. This chapter has provided a short introduction to the concept of metaphor, to the use of metaphor in design and to autonomous road vehicle metaphors.

This chapter introduced the seven key requirements of linguistic metaphors and noted that linguistic metaphors can be classified into three categories according to the nature of the comparison which is made: ontological, orientational and structural. And this chapter also introduced visual metaphors. It was noted that the images which are used can be arranged in three possible ways: juxtaposition, fusion and replacement. And that the underlying logic of the visual composition can be one of three types: connection, similarity or opposition.

Design involves a potentially unlimited number of physical, perceptual, cognitive and emotional characteristics which require specification. Tools which can help to simplify, select and communicate are thus extremely helpful. Metaphors are among the most important of those tools because they help to capture important concepts, goals and requirements.

Metaphors have been claimed to enhance creative processes such as design by supporting the emergence of new concepts from familiar ones. The use of common metaphors when working in a group can help to stimulate new ideas. And the concept of generative metaphor was introduced, a term which expresses how the framing and perspective which are provided by a given metaphor can help to bring new concepts into existence.

This chapter introduced the properties of salience and relatedness. Salience refers to the purity of the structure which the source of the metaphor offers, while relatedness refers to the overall conceptual closeness between the chosen metaphor source and the target artefact. It was suggested that selecting a highly salient and obviously related metaphor source renders the design process simpler, easier and more communicable.

The importance of metaphor when specifically designing interfaces was noted. With any product, system or service the user interface is the point of contact between the artefact and the human, and thus greatly influences the interactions and perceptions. Three measures of transfer from a metaphor source to a target interface were introduced: coherence, coverage and compliance.

It was suggested that early implementations of autonomous road vehicles will most likely improve upon and extend existing overall road vehicle metaphors, but that with time the power of automation will be harnessed to provide additional capabilities which diverge from our current conceptions of road transport. It was further suggested that the service which is provided will likely be at the heart of the future autonomous road vehicle metaphors, replacing to some extent the physical characteristics such as the on-road or off-road capabilities which were so dominant in the metaphors of 20th century human-driven vehicles.

From among the possible autonomous road vehicle metaphors which are currently being debated, four were selected as illustrative examples: autonomous shuttle, autonomous office, autonomous transport for use by specialist communities and autonomous entertainment centre. The aims, objectives and characteristics of each were briefly described.

This chapter noted that design metaphors are likely to prove important not just for the complete vehicle but also for individual components and subsystems. The dashboard of current human-driven vehicles was presented as an example of an automotive subsystem which is undergoing major changes to its metaphor at the moment as additional capabilities are added. And the route planning and route following subsystem of future autonomous road vehicles was cited as a subsystem which would benefit greatly from a single coordinating metaphor.

Finally, it was emphasised that the future friendly neighbourhood robots are likely to have fundamentally different subsystems for interacting with people and that each will need a design metaphor, either borrowed from a non-automotive environment or established ex novo, which can help to simply the design decisions and coordinate the characteristics.

Having introduced what is perhaps the best simplifying and focussing tool available to designers, that of metaphor, the next chapter will discuss instead the actual point of contact between people and friendly neighbourhood robots: interactions.

References

Beyer, A. 2016, Le transport fait-il partie du voyage? Pour une compréhension du déplacement touristique à partir de l'antagonisme contrainte/agrément, Géotransports, Vol. 7, pp. 7–22.

Black, M. 1955, Metaphor, Meeting of the Aristotelian Society, 21 Bedford Square WC1 London, May 23rd.

Carroll, J.M. and Mack, R.L. 1999, Metaphor, Computing Systems, And Active Learning, International Journal Of Human–Computer Studies, Vol. 51, No. 2, pp. 385–403.

Casakin, H.P. 2007, Factors Of Metaphors In Design Problem-Solving: implications for design creativity, International Journal Of Design, Vol. 1, No. 2, pp. 21–33.

Celentano, A. and Dubois, E. 2014, Metaphors, Analogies, Symbols: in search of naturalness in tangible user interfaces, Procedia Computer Science, Vol. 39, pp. 99–106.

Cila, N., Hekkert, P. and Visch, V. 2014, Source Selection In Product Metaphor Generation: the effects of salience and relatedness, International Journal Of Design, Vol. 8, No. 1, pp. 15–28.

Lakoff, G. and Johnson, M. 1980, Metaphors We Live By, The University of Chicago Press, Chicago, Illinois, USA.

Phillips, B.J. and McQuarrie, E.F. 2004, Beyond Visual Metaphor: a new typology of visual rhetoric in advertising, Marketing Theory, Vol. 4, No. 1–2, pp. 113–136.

Rojas, R. 2021, In The Heart Of Nashville, Rolling Parties Rage At Every Stoplight, New York Times, September 20th, www.nytimes.com/2021/09/19/us/nashville-party-vehicles.html

Schön, D.A. 1993, Generative Metaphor – a perspective on problem-setting in social policy, In A. Ortony (Ed.), Metaphor And Thought (pp. 137–163), Cambridge University Press, Cambridge, UK.

Schukert, M. and Müller, S. 2006, Erlebnisorientierung im touristischen transport am beispiel des personenluftverkehrs. In K. Weiermair and A. Brunner-Sperdin, Erlebnisinszenierung Im Tourismus (pp. 153–166), Erich Schmidt Verlag, Berlin, Germany.

Torgny, O. 1997, Metaphor: a working concept, Report CID-12, KTH, Stockholm, Sweden.

Turner, M. 1987, Death Is The Mother Of Beauty: mind, metaphor, criticism, University Of Chicago Press, Chicago, Illinois, USA.

Wayner, P. 2013, Future Ride: 99 ways the self-driving autonomous car will change everything from buying groceries to teen romance to surviving a hurricane to turning ten to having a heart attack to building a dream home to simply getting from here to there, CreateSpace Independent Publishing Platform, South Carolina, USA.

Chapter 8

Interaction

Interaction

Dictionary entries for the word "interaction" usually list at least three concepts:

- the transfer of energy between elementary particles, between a particle and a field, or between fields;
- a mutual or reciprocal action or influence;
- a situation where two or more things or people communicate with each other or react to each other.

The word "interaction" therefore involves either physical exchanges or symbolic (communicative) exchanges, or both. Whether in physics, psychology or sociology, interaction is a word used to describe situations in which something occurs between two or more entities, with each effecting the other to some degree, often as part of a sequence of communication.

Interaction has been defined by Crawford (2002) as a conversation between two actors who "listen, think and speak", emphasising that it involves give and take. It is not simply a reaction to something. An interaction is a dynamic sequence of events which is not fully controlled by one side or the other. It involves both sides, or possibly multiple sides when more than two objects or two parties interact.

With respect to products and services Rogers et al. (2019) have suggested five ways by which a user can interact with them:

- instructing: where users issue instructions to a system;
- conversing: where users have a dialog with a system;
- manipulating: where users interact with objects in a physical or virtual space by manipulating them (opening, holding, closing, placing, etc.);

DOI: 10.4324/9781003319740-8

- exploring: where users move through a physical environment or a virtual environment;
- responding: where the system initiates the interaction and the user chooses whether or not to respond.

One highly influential way of thinking about interactions was first introduced by the psychologist James Gibson (1966) who used the term "affordance" to describe perceptual features in the environment which facilitate interaction. The idea that things can perceptually announce their ability to perform certain interactions has over the years become widely accepted, and often provides a focal point for design. Saying that something exhibits an "affordance" is usually these days taken to mean that it has a shape, colour, sound or other perceptual properties which make it relatively easy to understand what it can do and what it can be used for. Some designed artefacts simply look like they wish to turn, or slide, or release or connect. They seem to demand a specific action from people.

Hartson (2003) extended the concept of affordance by considering in detail how the objects announce their capabilities to human observers. Hartson concluded that there were at least four types of affordance:

- physical affordance: design feature that helps users in doing a physical action;
- sensory affordance: design feature that helps users sense something;
- cognitive affordance: design feature that helps users know a state or sequence;
- functional affordance: design feature that helps users accomplish some task.

Hartson's approach extended the concept of affordance from a somewhat static set of physical and perceptual characteristics to a more dynamic understanding involving action, experiencing and learning. The concept of "affordance" was stretched from being mostly about communicating a capability from a distance to also communicating opportunities during use. With this definition in mind, any characteristic which attracts attention or any propensity for motion and action can be scrutinised to establish if it provides a useful affordance. And if not already present, helpful affordances can be chosen and designed for. Not designing affordances is not, cannot, be good design.

During any design project the materials, sizes, shapes, stiffnesses, masses, geometries, text, visuals, sounds and other parameters are usually carefully evaluated in terms of their appropriateness for the desired interactions. The artefact's interface in particular is usually the subject of much scrutiny, with the designers searching for affordances which can be designed in. If people can't interact with the artefact in effective, efficient and satisfying ways through the interface, then the commercial or societal success is at risk.

Achieving a safe, efficient, interactive and emotionally engaging interface is so critical to success as to merit a field of study dedicated specifically to the knowledge and skills involved, the field in interaction design. Fortunately, today, many useful interaction design methods and criteria are available.

Interaction Design

Early efforts at designing interactions usually focussed on the physical characteristics of the interface. The first mechanical and electrical devices were highly specialised pieces of equipment which were used mainly by trained professionals. Given the level of operator familiarity with the nature, parameters and processes of the task, the attention of the designers was usually directed at the efficiency of handles, wheels, knobs, buttons and other simple control surfaces. The focus was very much on achieving physical characteristics for the interface which facilitated the successful completion of the predetermined task. The paradigm at the time was that of "interface design".

Early interface design usually considered the interface to be those parts of the machine which the human operator could see, hear, touch or move. Thus the interface designers sought ways of improving the viewing, hearing, touching and moving. With time, ergonomic criteria emerged, such as the possibility of simplifying an interface and rendering it more intuitive by grouping controls of similar nature. Or by homogenising the control characteristics in order to achieve similar look, feel and operation. Control coding (Kroemer 2017) became a routine practice as designers attempted to group and homogenise according to:

- location: controls with similar functions can be located together;
- shape: control shapes can resemble the controlled component or controlled device;
- size: different sizes can be used to subdivide different groups of controls;
- mode of operation: the movement and resistance can be similar for controls which have similar functions;
- labelling: the same symbols can be used to identify controls which have similar functions;
- colour: red, orange, yellow or white can be used to signify special or safety-critical functions.

As the availability of complex machines became more widespread, however, it was noted with concern that the use of devices such as radar or the computer suffered errors which were attributable more to the decisions of the operator than to noise or malfunctions of the device. Stereotypes such as "pilot error" or the "distracted driver" are the residues of those early studies of human–machine interaction. A better understanding of the human was needed and was sought.

Awareness gradually set in that task performance was not only a function of the machine interface, but was instead an emergent property of the complete system composed of human and machine. Interface design based on the optimising of the control surfaces was simply not enough. The idea thus began to take hold that designers should consider matters from a systems perspective, thinking of the interactions between the human and the machine as a combined whole. The attention focussed increasingly on the dynamics of the exchanges. Human perceptual, cognitive and emotional characteristics came to be understood as design constraints which were every bit as important as the physical constraints which the control surfaces of the machine had to respect.

A step forward in the understanding and integration of the human perceptual, cognitive and emotional characteristics was achieved with ISO standard ISO/TR 16982: Ergonomics Of Human–System Interaction – usability methods supporting human centred design. The 2002 standard defined usability to be "the extent to which a product can be used by specified users to achieve specified goals with effectiveness, efficiency, and satisfaction in a specified context of use". The standard provided guidance about the most frequently noted issues of the day, and introduced methods to help designers better understand and better address usability requirements during design. Important points raised in the standard included the need to consider all the potential users, the need to consider the context of usage, the need to deploy multiple usability methods when evaluating and the need to incorporate actual user feedback into the design process.

The concepts of "usability" and of "user centred design" have become influential over the years and have expanded to cover a range of considerations beyond task performance. Rogers et al. (2019) suggested that "usability refers to ensuring that interactive products are easy to learn, effective to use, and enjoyable from the user's perspective". They noted six usability goals which can guide designers:

- safety;
- memorability;
- learnability;

- utility;
- effectiveness;
- efficiency.

The evolution in viewpoint which moved from "interface design" to "usability" then progressed onwards to the even more encompassing paradigm of "interaction design". The change in perspective was expressed well by Cooper (2004) in relation to digital artefacts when he stated: "I prefer the term interaction design to the term interface design because interface suggests that you have code over here, people over there, and an interface in between that passes messages between them. It implies that only the interface is answerable to the user's needs. The consequences of isolating design at the interface level is that it licenses programmers to reason like this: I can code as I please because an interface will be slapped on after I'm done. It postpones design until after the programming, when it is too late."

The term "interaction design" has often been associated with the 1980s work of Bill Moggridge, the cofounder of IDEO. He championed the empirical exploration of interactions and the fitting of new technologies to the characteristics and abilities of humans. Four basic dimensions of interaction design were defined in his well-known book *Designing Interactions* (Moggridge 2007):

- words: are the elements of an interface that users interact with;
- visual representations: are the elements of an interface that the user perceives, which may include typography, diagrams, icons, and other graphics;
- physical objects or space: the objects with which the users interact and the space within which the interaction takes place;
- time: the time during which the user interacts with the interface.

Also adopting an interaction design perspective, Ben Shneiderman suggested eight golden rules for interfaces (Shneiderman et al. 2016):

- strive for consistency;
- enable frequent users to use shortcuts;
- offer informative feedback;
- design dialog to yield closure;
- offer simple error handling;
- permit easy reversal of actions;
- support internal locus of control;
- reduce short-term memory load.

The arrival of increasingly more complex computers and automation has exposed, however, the limits of the practical considerations articulated by professionals such as Moggridge and Shneiderman. While the sensory modalities remained much the same, the complexity and richness of the interactions has increased. In response, interaction designers have had to dig more deeply into the psychological and sociological considerations involved.

Rogers et al. (2019) have suggested that interaction design has become the practice of "designing interactive products to support the way people communicate and interact in their everyday and working lives". Cooper et al. (2014) stated matters even more bluntly as "If we design and develop digital products in such a way that the people who use them can easily achieve their goals, they will be satisfied, effective and happy". Such proposals allude to a depth and richness of the design considerations which go well beyond the functional aspects of the interfaces. Interaction design is in fact now a complex multidisciplinary activity involving a variety of criteria and techniques to shape the exchanges between humans and machines.

And the public awareness of the importance of such matters has also grown. For example, people have increasingly found that the achievement of their goals and the possibilities for happiness in their lives depend critically on the screens and input controls of today's ubiquitous digital devices. What had at one time been a mostly functional matter of "interface design" is today a much more meaningful matter of "interaction design" due in large part to the devices having permeated nearly every aspect of people's public and private lives. Continuous exposure to devices has amplified some traditional concerns of human–machine interaction and has led others to grow into priority areas.

Two concerns have been chosen to illustrate here the variety of issues and meanings involved. The two were selected based on their importance in relation to people's goals and happiness. And each involves functional and ethical considerations which tend to be a major preoccupation of designers today. The two are: frictionless interaction and algorithmic transparency.

The concept of "frictionless interaction" is a priority in the design of many of today's products, systems and services, particularly those which are primarily digital in nature. Frictionless interaction refers to an interaction whose physical, perceptual, cognitive and emotional characteristics are so intuitive and natural to most people as to provide a negligible step on the way to achieving some goal. The concept was popularised by the book *Don't Make Me Think!* (Krug 2000) whose premise was that a software programme or website should help users accomplish their tasks as quickly and easily as possible, with minimal mental effort.

Achieving frictionless interaction requires the use of components and conventions which follow as much as possible the instinctive and intuitive behaviours of people. The word "frictionless" is thus a design sector shorthand for saying "which affords natural human patterns of thought and motion". Frictionless interaction requires the maximum possible simplification of all visuals, sequences and selections and the elimination of unnecessary or unfamiliar objects and items of information. Where the interactions are mainly cognitive in nature the concept of "frictionless" is often taken to mean simply "low cognitive workload". For the designers of many internet websites the priority is to keep things as simple and obvious as possible so as to reduce the cognitive workload.

Another priority in the design of many of today's products, systems and services is that of "algorithmic transparency". The terms "transparency" or "interpretability" are often used in design discussions to indicate the user's understanding of what the artefact is actually doing. With the simple mechanical machines of the past there was usually little difficulty in understanding the settings or the operating state of the machine. Simple viewing windows, indicators or dial gauges were usually sufficient to convey the key items of information which might be of interest to the human operator or to other interested parties. The situation is however different with the complex multifunction automated machines of today.

The complexity introduced by automation has greatly increased the difficulty of providing obvious, intuitive, transparent and interpretable responses to any given human input. As the decision making on the part of the automation has increased, and the sophistication of its functions widened, the human understanding of what is happening has usually suffered. Whereas there can be little doubt regarding what is happening when working with a hammer or wrench, doubts and confusions can easily occur when interpreting the findings of a Google search or when attempting to understand an automobile engine malfunction.

The expression "algorithmic transparency" was popularised by Diakopoulos and Koliska (2017) in relation to the understanding of the content selection decisions made by digital journalism services. Their research considered how users can be informed of the selection criteria and processes, and what effects such knowledge might have towards public opinion forming. In recent years the use of the term has however spread, and it currently refers to the knowability of the automation's decisions and actions in a general sense.

One approach for improving "algorithmic transparency" is the use of explanatory statements (Friedrich and Zanker 2011) during the interactions. Rader et al. (2018) have suggested the possibility of at least four forms of explanation which can help to improve the algorithmic transparency:

- how explanations: descriptions of a system's inputs and outputs and the steps it takes to arrive at a particular outcome;
- why explanations: justifications for a system and its outcomes and the explaining of the motivations behind the system;
- what explanations: descriptions which reveal the existence of algorithmic decision making, but without providing additional information about the system;
- objective explanations: descriptions of the process by which a system comes into being and is continually improved.

And a relatively comprehensive approach for addressing algorithmic transparency is provided by the recently approved IEEE standard P7001-2021 (Winfield et al. 2021). It establishes five target audiences: users, general public, certification agencies, incident investigators and administrative/litigation experts. And for each of the target audiences the requirements are organised into five levels of increasing transparency. Across the ensemble of audiences and levels can be found requirements such as user manuals, data logging systems, visual information systems, natural language interfaces and request functions which can be used to interrogate the system. Though not addressing the issue of operating context or the strengths and weaknesses of individual items of information, the standard nevertheless provides a framework for design.

While not fully separable from the issue of anthropomorphism, algorithmic transparency has nevertheless become an area of research in its own right. People are increasingly expecting immediate insights into motivations, processes and outcomes rather than relying on slower and more traditional approaches such as user manuals, internet tutorials or telephone helplines. Increasingly detailed guidelines are thus being proposed for the revelatory strategies, and an increasing number of tests of human comprehension are being devised. Given the obvious ethical implications of unclear or misleading items of information, it is rather likely that the matter of algorithmic transparency will also one day end up within the formal realm of corporate or governmental regulation.

Interaction Measurements

Humans seem to like measurements. Particularly numbers. Archaeologists have found artefacts from more than 40,000 years ago with cuts which appear to be tally marks made by carving notches in the wood, bone, or stone. Some have been suggested to be counts of days, or lunar cycles, or records of important possessions such as animals. Measuring things, and keeping a record, appears to be as old as humanity itself.

It is therefore unsurprising that quantitative evidence (numbers) is often preferred to qualitative evidence (judgements or ideas) when designers are asked to provide evidence of the performance of their interfaces and interactions. It is difficult to escape the urge to measure, despite many human interactions not easily lending themselves to measurement.

The multiplicity of interactions with complex machines and the many goals, thoughts and actions involved suggest that any attempt at measurement will provide only a partial picture. There are simply too many ways in which interface components or the items of information involved might stimulate memories and thoughts. And the more complex the activity, the more possibilities there will be for distractions and misunderstandings. Nevertheless, partial pictures can be helpful, and are often a required part of formal validation.

One of the earliest forms of measurement which involved an interface was that of the receiver operating characteristic (ROC) which was developed for quantifying human detection capabilities (Swets 1996). With the birth of electronic tools such as radar and sonar it was observed that the operator's ability to notice a signal on the screen or through the headphones was a complex function of some of the properties of the electronic device (for example the display type, display size, display intensity, etc.) and some of the properties of the human operator (for example the time of day, workload, fatigue state, etc.). While not strictly a measure of interaction, ROC curves were nevertheless an early step towards considering interactions as "measurable".

In the 1960s the growing complexity of power systems, military vehicles, aircraft and spacecraft stimulated measurements of the key interactions of their operators. And by the 1970s the concept of workload, or, more specifically, task workload, became an important consideration. It refers to the relationship between the amount of available human mental processing capability and the amount actually needed for the task which is being performed. If a machine operator feels that a given task is easy, requiring only a small amount of available mental capacity, then the task workload is said to be low. The concept proved useful for understanding human performance and proved particularly helpful in the assessment of situations where inadequate interface characteristics stimulated safety-critical errors.

By the 1980s techniques such as the NASA Task Load Index (TLX) were developed (see Figure 8.1) to measure the physical and mental

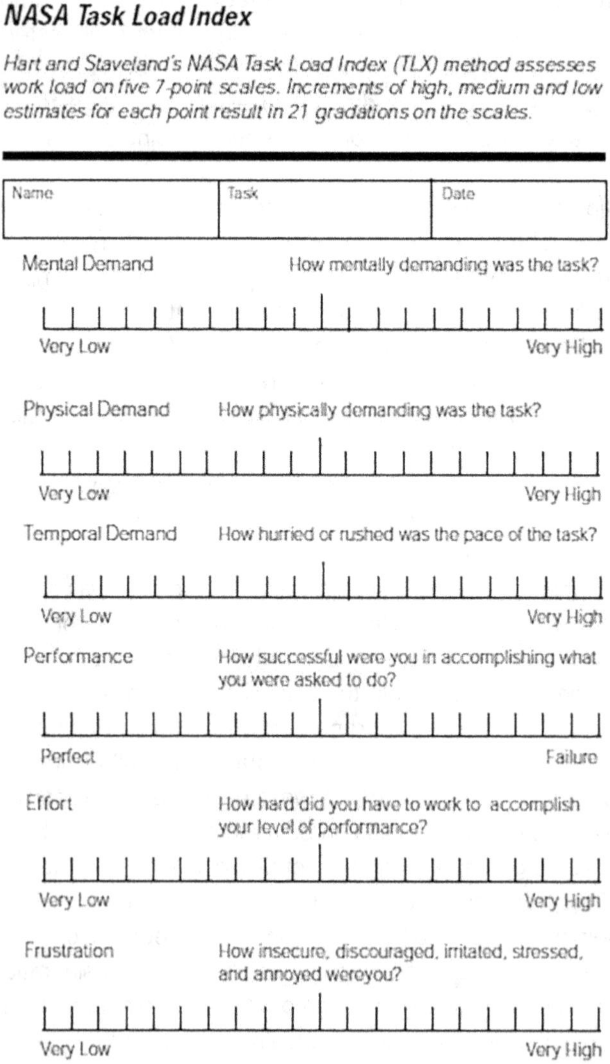

NASA Task Load Index

Hart and Staveland's NASA Task Load Index (TLX) method assesses work load on five 7-point scales. Increments of high, medium and low estimates for each point result in 21 gradations on the scales.

Name	Task	Date

Mental Demand   How mentally demanding was the task?

Very Low   Very High

Physical Demand   How physically demanding was the task?

Very Low   Very High

Temporal Demand   How hurried or rushed was the pace of the task?

Very Low   Very High

Performance   How successful were you in accomplishing what you were asked to do?

Perfect   Failure

Effort   How hard did you have to work to accomplish your level of performance?

Very Low   Very High

Frustration   How insecure, discouraged, irritated, stressed, and annoyed were you?

Very Low   Very High

Figure 8.1 The NASA Task Load Index.

Source: Wikimedia Commons

workload arising from the use of an interface. While not strictly a measurement of "interaction", the NASA TLX (Hart 1986) does nevertheless furnish information about the interaction and provides a neutral language which can be used to compare different interfaces, different contexts, different times or different time durations. Some additional activity and designer interpretation of the results is necessary to identity the exact elements of the machine interface which caused the feelings such as "effort" or "frustration", but the NASA TLX nevertheless has proved helpful for the assessment of many safety-critical systems and is still widely used today.

In the 1980s the growing complexity of machines also led to the concept of situation awareness (Endsley and Jones 2012). Endsley (1995) defined situation awareness as "the perception of environmental elements and events with respect to time or space, the comprehension of their meaning, and the projection of their future status". Situation awareness is a term which refers to a person's understanding of the things which matter in the surroundings, whatever "matter" might mean in the given context at the given point in time. For example, when approaching a road junction an automobile driver might feel that things like the vehicle's speed, the distance to the junction and the presence of other vehicles at the junction might "matter". Situation awareness is an important consideration for complex tasks such as driving an automobile, affecting response times, decision making and safety.

Several techniques were developed to estimate situation awareness in an indirect or a direct manner (see Endsley and Jones 2012 for a review). Among the indirect techniques are the recording of the operating states of the machine, verbal protocol analysis, communication analysis (such as between pilot and control tower), physiological measures such as electrocardiograms (ECG) and electroencephalograms (EEG), behavioural measures and recording of the process outcomes.

The so-called direct estimation of situation awareness involves instead asking the operator for judgements or self-ratings about his or her understanding of the situation using predetermined criteria which are considered relevant. The opinions are usually recorded at the end of an activity involving either a physical or a simulated version of the system, and the activity usually involves some predetermined task so as to facilitate comparisons between the ratings expressed by different people or for different versions of the system.

Among the most commonly used of the direct estimation methods is the Situation Awareness Rating Technique (SART). It is a post-trial subjective evaluation technique originally intended for the assessment of aircraft interfaces (Taylor 1990). As shown in Figure 8.2 the SART

Instability Of Situation.
How changeable is the situation? Is the situation highly unstable and likely to change suddenly or is it very stable and straightforward?

Low	1	2	3	4	5	6	7	High

Variability Of Situation
How variable is the situation? Is it complex with many interrelated components or is it simple and straightforward?

Low	1	2	3	4	5	6	7	High

Complexity Of Situation
How complicated is the situation? Is the situation highly unstable and likely to change suddenly or is it very stable and straight forward?

Low	1	2	3	4	5	6	7	High

Arousal
How aroused are you in this situation? Are you alert and ready for activity or do you have a low degree of alertness?

Low	1	2	3	4	5	6	7	High

Spare Mental Capacity
How much spare mental capacity do you have in this situation? Do you have sufficient capacity to attend to many variables or nothing to spare at all?

Low	1	2	3	4	5	6	7	High

Concentration
How much are you concentrating on the situation? Are you concentrating on many aspects of the situation or focussed on only one?

Low	1	2	3	4	5	6	7	High

Division Of Attention
How much is your attention divided in this situation? Are you concentrating on many different aspects of the situation or focussed on only one?

Low	1	2	3	4	5	6	7	High

Information Quantity
How much information have you gained about the situation? Have you received and understood a great deal of information or is it very stable and straightforward?

Low	1	2	3	4	5	6	7	High

Information Quality
How good is the information which you have gained about the situation? Ids the knowledge communicated very useful or is it insufficient?

Low	1	2	3	4	5	6	7	High

Familiarity
How familiar are you with the situation? Do you have a great deal of relevant experience or is this a new situation?

Low	1	2	3	4	5	6	7	High

Figure 8.2 The Situational Awareness Rating Technique (SART).

consists of ten questions which cover the three general areas of attentional demand, attentional supply and understanding. Asking people to express their opinions via SART ratings provides a standardised basis for comparison of different designs and permits a degree of cross comparison to related systems and to other operational environments.

Another commonly used direct estimation method is the Situation Awareness Global Assessment Technique (SAGAT). It is based on the freezing of the activity or mission simulation at randomly selected points in time, then blanking the system displays and asking the person questions about the situation (see Endsley 1988 for details). The idea behind the SAGAT is to compare the person's understanding of the operational environment to the actual conditions at multiple points in time. The answers provided by the person are compared to the actual situation so as to provide an objective measure of the situation awareness. The greater the deviation of the person's understanding of the operational environment, the less effective the situation awareness system is considered to be.

While most of the early interaction measurements were for military, aeronautical and space systems, performing them for road vehicles also became common. Automotive systems became more complex as new displays and input devices were added, often leading to difficult interactions and safety risks. Road vehicle designers therefore found themselves grappling with many more interactions than in the past, and those interactions were often of noticeably different nature.

The proliferation of information, decisions and interactions eventually persuaded automotive designers to seek inspiration from the world of computers. Being machines which were invented for the purpose of executing lengthy lists of complex instructions, computers perform sophisticated tasks which lead to a variety of interaction design issues of a sensory or cognitive nature. Many measurements of interaction were thus devised for use with computers, and automotive designers have increasingly borrowed those metrics as part of their efforts to achieve better vehicles. Much has been written in the press in recent years about "computers on wheels", capturing the general feeling that the centre of gravity of automotive design has been shifting. And automotive interfaces such as those of today's infotainment systems do in fact seem closer to computer interfaces than to the dashboards of yesteryear.

Metrics of human–computer interaction can be grouped into three general categories (Tullis and Albert 2013): performance metrics, physiological metrics and self-evaluated rating metrics. None of them can be considered to provide a definitive, scientific, evaluation of the effectiveness or efficiency of an interaction. All, however, provide items of information which can assist designers by providing evidence of the effects of their design choices.

The performance metrics capture the direct effects of the interaction. Since for example the concept of "usability" refers to the ease with which a person achieves a predetermined goal, it makes sense to measure whether the goal was actually achieved or not. And measuring how long it took to achieve the goal or how many steps the process involved would also seem to be useful. Measuring a few things at the machine end of the interaction provides undeniable facts about what occurred.

Tullis and Albert (2013) suggested five types of performance metric:

- task success: the measurement of the completion of a given task either in absolute (yes/no) terms or in terms of degree (based on how much or how closely the actual outcome resembles the target outcome);
- time on task: the measurement of the amount of time taken to complete the given task;
- errors: the measurement of how many incorrect task completions occurred;
- efficiency: the measurement of the effort required to complete a given task, in terms of the number of button presses, click selections or other actions;
- learnability: the measurement of the change in user behaviour over time, usually by comparing one or more of the other performance metrics at different points in time.

Of the three categories of human–computer interaction metrics, that of the performance metrics can be considered to be the most scientific and the most objective. Performance metrics are typically numbers which quantify physical matters such as machine states or time durations. They are relatively unambiguous items of information.

Physiological metrics involve instead the measurement of things at the human end of the interaction. Several medical and biomedical techniques can be used to identify changes in people's bodily parameters in response to an interaction with a designed artefact such as a computer. While frequently used, and often helpful, the physiological metrics can be somewhat harder to interpret and can sometimes be affected by matters which have little or nothing to do with the interaction. A person's heartrate can for example rise or fall due to mental activities which are not related to the interaction which is being investigated, such as thinking about a loved one.

Commonly used physiological metrics include skin conductance, heart rate and heart rate variability. Complete electrocardiograms (ECG) are also sometimes analysed in search of higher order statistical changes in pattern. Electroencephalograms (EEG) have been used in some research studies with LPP, P300 or the statistics of the delta (0–4 Hz), theta (3–7 Hz), alpha (8–12 Hz), beta (13–30 Hz) or gamma (30–40 Hz) wave bands serving as indicators of attentional focus and neural processing. The assumption underlying the use of all such metrics is that the interaction with the artefact is likely to produce a change in the physiological signal which is being monitored. Changes in the physiological signals are therefore suggestive of some change in the nature or the intensity of the interaction, and systematic changes in one direction or the other can be indicative of improving or worsening interaction.

Among the more recent physiological metrics is that of eye tracking (Bergstrom and Schall 2014). Eye tracking uses infrared light sources (emitters) and infrared video cameras (detectors) to measure where the person is looking at each instant in time. With eye tracking the location being viewed, and the statistics describing how often or how long the person views, provide metrics of what a person is paying attention to and presumably thinking about. These days eye-tracking is routinely used when designing computers, websites and other artefacts which necessitate substantial visual processing.

And, most recently, physiological metrics have been developed which are based on facial or vocal emotion analysis (Meiselman 2016). Expert systems and deep learning neural networks have permitted the automation of the process of detecting and classifying expressions of human emotion, from either a camera image of the person's face or from microphone capture of the person's voice. The assumption underlying the use of emotion analysis is that the interaction with the artefact is likely to stimulate an emotion of one type or other, to some degree of intensity. And that these physiological responses are suggestive of the interaction's nature and intensity, with positive emotional responses usually being considered preferable to negative emotional responses.

Facial expression analysis (Akamatsu 2019) provides automated implementation of the ideas which were first popularised in the 1960s by Ekman (Ekman and Friesen 2003). Current systems usually provide real-time measurements of the six originally defined basic emotions (fear, anger, distress, disgust, surprise and joy) and often provide additional measures such as one or more of the higher order cognitive emotions (pride, jealousy, envy, shame, embarrassment, guilt and love).

Vocal emotion analysis instead uses pattern recognition to detect emotions from acoustic signals. The analysis is complicated by the fact that human sounds are a mixture of evolutionarily ancient sounds which express affect, and the evolutionarily more recent sounds of words which can sometimes also refer to affect. The extent to which the non-verbal and verbal sounds can be separated is a subject of ongoing debate, however, several commercial vocal emotion analysis technologies are now in use. While academic research (Cowen et al. 2019) suggests the possibility of expressing 24 or more emotions via the human voice, current technologies provide a smaller subset consisting usually of seven key emotions (anger, disgust, fear/panic, happiness, sadness/sorrow, boredom and neutral).

While the physiological metrics of human–computer interaction are scientific measurements, it is nevertheless the case that they are indirect in nature and can thus be easily affected by metabolism, body movements or other confounding factors. They quantify human responses such as heart rate, eye position or emotional expression, but the numbers and statistics can obscure the fact that many assumptions and simplifications are needed for those numbers to have meaning in relation to interactions. Concluding that one form of interaction is better than another because the heart rate had dropped across a group of participants is a risky business at best.

Of the three categories of human–computer interaction metrics, that of the self-evaluated ratings can be considered to be the most subjective and most personal. Metrics in this category have the common characteristic of being explorations of the person's perceptions and feelings about the interaction. They come in a variety of forms and adopt a variety of measurement scales. Review of the linguistic and psychological guidelines in relation to question formulation are beyond the scope of this current book. Also beyond the scope of this book is discussion of the strengths and weaknesses of the Likert, Semantic Differential and other commonly used scales. The questions and formats are the subjects of several well-known texts (see for example Hayes 1992 and Tourangeau et al. 2000) and the reader is directed to those authoritative sources for details.

Only two approaches are discussed here. They illustrate the variation in psychological focus and measurement scale involved in evaluating complex systems such as autonomous road vehicles. Each is considered by the author to be representative of autonomous road vehicle interactions which need to be evaluated in some manner in some situations. While experience suggests the usefulness of also several other well-known metrics, space considerations limit the current exposition to the two illustrative examples.

The first example is the System Usability Scale (SUS) which provides a global overview of perceived usability. SUS was originally developed to evaluate electronic office systems (Brooke 1996) thus it is best suited to predefined tasks which are reasonably well understood by the user. To achieve balance and reduce bias, half of the questions (see Figure 8.3) are worded negatively and half are worded positively. Statistical analysis of the results is facilitated by the use of numbered Likert Scales (see Hayes 1992) for the responses.

SUS is a popular approach which has a long history of use in diverse areas of design, providing opportunities for benchmarking against existing artefacts which share similar characteristics or similar goals. Its generality and simplicity make SUS an ideal candidate for evaluations of different designs or of the same design in different environments or different social contexts.

Methods such as the SUS can be helpful with autonomous road vehicles because there are some interactions which are highly functional in nature. Interactions such as those needed for the climate control are characterised by somewhat standardised requests and somewhat standardised items of information. In some cases a single number, the desired temperature, may suffice to adequately instruct the vehicle. Such interactions come close to constituting well-defined and reasonably well-understood tasks of the type which the SUS was originally developed for.

But, unfortunately, not all autonomous road vehicle interactions will be well defined tasks which are reasonably well understood by the user. Some autonomous road vehicle interactions will instead be closer to the conversational and social interactions which regularly occur between humans. With passengers who are not specially trained operators, a variety of gestures, words and phrases may end up being used to indicate road dangers to the vehicle or to request services from the vehicle. And those gestures, words and phrases may vary significantly from person to person. Choosing an optimal background music for the journey, or the optimal interior lighting, or some other non-scientific matter may prove more an exercise in conversation and social interaction than a well-defined task.

The second example discussed here is the Naturalness Of Interaction Scale proposed by Ramm (2018) for use specifically in automotive design (see Figure 8.4). It is a more recent development which addresses some of the issues which can arise when the interaction is not about a simple, fully defined and fully familiar task. It may prove useful for evaluating some of the more conversational and social interactions which are inevitable with some autonomous road vehicles.

I think that I would like to use this system frequently.	Strongly Disagree	1	2	3	4	5	Strongly Agree
I found the system unnecessarily complex.	Strongly Disagree	1	2	3	4	5	Strongly Agree
I thought the system was easy to use.	Strongly Disagree	1	2	3	4	5	Strongly Agree
I think that I would need the support of a technical person to be able to use this system.	Strongly Disagree	1	2	3	4	5	Strongly Agree
I found the various functions in this system were well integrated.	Strongly Disagree	1	2	3	4	5	Strongly Agree
I thought there was too much inconsistency in this system.	Strongly Disagree	1	2	3	4	5	Strongly Agree
I would imagine that most people would learn to use this system very quickly.	Strongly Disagree	1	2	3	4	5	Strongly Agree
I found the system very cumbersome to use.	Strongly Disagree	1	2	3	4	5	Strongly Agree
I felt very confident using the system.	Strongly Disagree	1	2	3	4	5	Strongly Agree
I needed to learn a lot of things before I could get going with this system.	Strongly Disagree	1	2	3	4	5	Strongly Agree

Figure 8.3 The System Usability Scale (SUS).

	Attribute A	Very A	Somewhat A	Neither A nor B	Somewhat B	Very B	Attribute B
Imagining the car is a person, the system seems:	Unhelpful						Helpful
Imagining the car is a person, the system seems:	Rude						Polite
The system seems:	Highly Incompetent						Highly competent
The car responds:	Unpredictably						Predictably
When you do the action it feels like:	The car is fully in control						You are fully in control
Operating the control feels:	Difficult						Easy
Mentally the interaction is:	Highly demanding						Not at all demanding
The interaction overall feels:	Counter intuitive						Intuitive
The communication between you and the car feels:	Artificial						Real
The control's response feels:	Delayed						Instantaneous
The car comes across as:	Uncommunicative						Informative
Overall the interaction felt:	Unnatural						Natural
The control is located:	Illogically						Logically
The shape and movement of the control:	Does not reflect its function at all						Closely reflects its function
The input action required seems:	Completely unclear						Completely obvious

Figure 8.4 The Naturalness Of Interaction Scale.

The scale is based on the results of contextual interviews and workshops which investigated the naturalness of interaction between people and automobiles. Adopting a semantic differential approach, some elements evaluate aspects of the interaction which are more mechanical in nature while others evaluate aspects of the interaction which touch upon transparency, agency and personality. The scale involves a mixture of metrics which span a range of considerations which are relevant to the complex systems which can be found in modern automobiles. Some of the questions are obviously appropriate to simple mechanical devices while others are better suited to the evaluation of complex decision-making machines.

The Naturalness Of Interaction Scale is simple and easy to use, thus like the System Usability Scale it provides opportunities for evaluating different autonomous road vehicle designs or the same autonomous road vehicle in different environments or different social contexts. It furnishes information about the interaction and provides a somewhat neutral language which can be used for comparisons. Autonomous road vehicles will have many interactions which are somewhat ill defined, conversational or social in nature, thus it seems likely that methods such as the Naturalness Of Interaction Scale will be needed for performance evaluation.

Interactions With Autonomous Road Vehicles

Figure 8.5 presents an image of a human driving a recent production automobile. Inspection of such road vehicles suggests a wealth of well-established visual, acoustic and tactile interactions which involve the

Figure 8.5 A human driving a recent production automobile.
Source: haiza3050

primary controls (steering, gearshift, etc.), secondary controls (wipers, heating, etc.), instruments (speedometer, fuel gauge, etc.), mirrors, navigator and infotainment system. After more than 100 years the choice of displays, choice of controls, positioning and physical characteristics have all been optimised.

Inspection of current road vehicles also suggests the wealth of well-established affordances. Examples include rocker switches for simple on–off controls, windshield wiper control stalks which resemble the wiper itself and which move in the same rotary manner, and digital screens which adopt familiar conventions from the world of computers. And safety implications have constrained designers to adopt primary controls (steering wheel and pedals) which are relatively pure affordances so as to optimise the speed, directness and efficacy of interaction. After more than 100 years the human driver is now well catered for in terms of interface affordances.

The situation is however different with autonomous road vehicles. At the time of writing of this book it is not yet possible to discuss autonomous road vehicle interactions with confidence. The reason is simple, they are not yet a commercial reality and thus their definitive characteristics are still largely unknown. As with the design metaphors which were discussed in the previous chapter, only speculations can be made at the current point in time. And those speculations can only be based on facts which are likely to remain somewhat stable going forward in time.

Key facts which are likely to remain stable going forward include the vehicle systems which are likely to need interactions. Key facts also include the operational design domain of the vehicle which prescribes and certifies its usage. And key facts include the characteristics of the likely services which the vehicle will provide.

Table 8.1 presents a list of nine technical subsystems which are likely to be part of future autonomous road vehicles which transport people. While all nine will be integrated together in some manner, each will also have specific functional and behavioural capabilities which lead to some separate interactions. Each will have to ensure that its interactions are appropriate and acceptable for people.

The design of several of the subsystems is likely to be based in part on traditional automotive practice. For example, the climate control, ingress/egress and breakdown and emergency subsystems of autonomous road vehicles will surely resemble those of current vehicles in many ways, and should thus prove somewhat familiar to people psychologically.

Others from among the nine subsystems will instead be based on new technologies. Those subsystems will have fewer precedents from past practice thus people may be prone to misunderstandings and

TABLE 8.1 Autonomous road vehicle subsystems.

System	Components
Ingress/Egress	door handles/interfaces doors door motors and actuators seats seat motors and actuators
Exterior Communications	front, rear and side body shapes exterior lighting systems and exterior displays acoustic warnings and acoustic communication systems
Interior Communications	infotainment system interior lighting systems and interior displays acoustic warnings and acoustic communication systems
Parking	haptic, manual, gesture-based and acoustic interaction systems situation awareness systems
Driving	haptic, manual, gesture-based and acoustic interaction systems situation awareness systems
Climate Control	haptic, manual, gesture-based and acoustic interaction systems acoustic warnings and acoustic communication systems window handles/interfaces windows window motors and actuators internal intercompartment aperture systems situation awareness systems
Refuelling/ Recharging	haptic, manual, gesture-based and acoustic interaction systems situation awareness systems acoustic warnings and acoustic communication systems contact and/or contactless power couplings
Internet, Social Media and Personality	haptic, manual, gesture-based and acoustic interaction systems emotion interaction systems digital agent(s)
Breakdown and Emergency	haptic, manual, gesture-based and acoustic interaction systems cellular and internet network connectivity systems digital assistance agent(s) human assistance agent(s)

errors during early encounters. Subsystems such as exterior communications, parking and driving will be substantially different from existing approaches because the vehicle will be performing the activities independently, rather than relying on a dedicated human driver. Much still remains to be learned about the psychological and sociological implications of the self-driving.

Regarding the underlying technical requirements, several research teams have put forward proposals for driving scenarios to evaluate. One proposal from the UK Transport Systems Catapult (Transport Systems Catapult 2017) suggests the set of "challenging driving scenarios" which is summarised in Table 8.2. The scenarios are driving situations whose complexity can prove challenging to current automation. They thus constitute a minimum set of driving conditions which should be evaluated to check the safety and possibly also the comfort of the autonomous road vehicle. Establishing possible ways of dealing with such driving situations is a first step in design, which will inevitably need to be followed by human evaluations of the frictionlessness, transparency, naturalness and intuitiveness of the driving actions.

Another example of proposed driving scenarios is the Safety Pool Scenario Database developed by the University of Warwick WMG (Warwick Manufacturing Group). The driving and emergency scenarios are intended for use in either numerical simulation or physical testing, and already include more than 100,000 situations which can prove challenging for driving automation. The standardised scenarios help bring into focus the effects of design decisions regarding the vehicle dynamics, sensors, signal processing algorithms, detection algorithms, mapping software, real-time communications, route-planning algorithms and control algorithms. Again, such technical tests will inevitably have to be followed by human evaluations of the frictionlessness, transparency, naturalness and intuitiveness of the driving actions.

Other key facts which will shape the interaction design of autonomous road vehicles are those which can be extracted from the vehicle's operational design domain (ODD). SAE standard J3016 (2018) defines the operational design domain to be the "operating conditions under which a given driving automation system or feature thereof is specifically designed to function, including, but not limited to, environmental, geographical, and time-of-day restrictions, and/or the requisite presence or absence of certain traffic or roadway characteristics".

The ODD is a mostly technical description of what the autonomous road vehicle is designed and certified to do, thus it provides the underlying technical requirements against which the design activity must be performed. The ODD is intended to include the main functional

TABLE 8.2 UK Transport Catapult list of challenging driving scenarios for autonomous vehicles (AVs).

Category	Abnormal / Challenging Driving Event	Issues Involved
Obstructions	Parked vehicles	How to ensure they are parked and have not momentarily stopped. How to allow for possibility of doors opening?
	Disabled (broken down / crashed) vehicles	Passing may lead to compromising other rules of road such as crossing solid white lines.
	Pedestrians	How much space to leave? Different clearances for different types? Pedestrian behaviour can be unpredictable. Should vehicle slow down when passing pedestrians on footway?
	Passing cyclists	How much space to leave? Different clearances for different types? Cyclist behaviour can be unpredictable.
	Road flooding	Difficult to sense the depth? Could lead to loss of control of vehicle or splashing of pedestrians.
	Animals in road (either shepherded or loose)	For smaller animals, it can be difficult to decide whether to pass over animal or attempt to stop or swerve.
	Ridden horses	Determining appropriate speed and overtaking strategy.
	Negative obstructions such as pot holes or road / bridge collapse	Could be difficult to sense.
	Load shedding from other vehicles	Action could depend on density / mass of objects being shed, but might not be possible for machine to determine.
	Vehicles in process of becoming disabled, e.g. tyre blow out, lorry jack-knifing, tall vehicle overturning in wind etc.	Challenging for machine to detect and interpret subtle clues that provide indications.
	Traffic calming measures	Speed humps, chicanes, etc. need to be detected and negotiated appropriately.
	Fallen power cables / branches in carriageway	Could be challenging to detect with sensors.
	Level crossing	Similar to traffic signal but consequences of stopping on rail line could be catastrophic.
	Overtaking	Challenging to detect oncoming vehicles.

Category	Item	Description
Lane reallocation/ rerouting	Temporary lane closure on highway	Road layout may differ from map being referred to by vehicle.
	Temporary contraflow	Automated driving feature may be designed for highways and not for two-way traffic operation.
	Lane designations (e.g. bus lanes, high-occupancy vehicle lane, hard shoulders)	May need to clarify under what circumstances vehicles can enter.
Adverse weather / environmental conditions	High winds	Loss of control.
	Snow either falling or on carriageway	Sensor visibility compromised, road markings and kerb lines obscured, loss of vehicle control.
	Heavy rain	Sensor visibility compromised, road markings obscured, loss of vehicle control.
	Ice	Loss of control.
	Fog / bright sunshine etc.	Sensor visibility compromised.
Road etiquette	Emergency vehicle in vicinity	How to avoid impeding whilst obeying rules of road.
	Crossing white lines	Under what circumstances can vehicle do this?
	Interpreting gestures from other road users	Challenging to detect and interpret the meaning of hand gestures, flashing of headlights, etc.
Traffic flow arbitration	Police or authorised persons intend to stop AV	How does AV recognise what is an authorised person, and then interpret commands?
	Two lanes merge into one	Often involves interaction between human drivers.
	Merging onto highway	Often involves interaction between human drivers.
	T junction	Poor lateral field of view from AV.
	Cross-roads	Poor lateral field of view, right turn stale-mate.
	Temporary speed limits	How to ensure location is communicated to AVs.
	Temporary traffic signals	How to ensure location is communicated to AVs.
	Temporary stop–go sign	Can AV interpret?
	Giving way to oncoming vehicles on narrow section of road	Often required communication between human drivers to decide who proceeds first.
	Roundabouts	Detecting correct lane allocations, "give way to right" standoff.
	Zebra crossings	Giving way to waiting pedestrians.
	Traffic signal failure	Junction reverts to priority based, or interaction between human drivers.

capabilities which are expected of the vehicle, stated at an intermediate-to-high level of abstraction. Examples of functional capabilities which can be part of an ODD of an autonomous road vehicle include those of the "interstate pilot" and the "autonomous valet parking" suggested by Wachenfeld et al. (2016). Such descriptions provide the goals which are at the heart of the interactions.

An ODD specification inspired by a metaphor such as "autonomous valet parking" provides some of the informatic and functional requirements of the intended interaction between the human and the vehicle, however such specifications are not detailed or prescriptive in human terms. ODDs can clarify, for example, that an autonomous road vehicle should park itself, but they do not provide guidance regarding how the vehicle should respond to specific requests and do not provide guidance about the physical, linguistic or temporal dynamics of the interaction. The sensory modalities, language requirements, tone, emotions or exact actions involved are not usually defined at present. Detailed information of a psychological or sociological nature is conspicuous by its absence.

And looking beyond the ODD, what should be the service design targets? What should be the response times, error rates, punctuality, degree of agency or level of politeness? Psychologically, 20th century automobiles provided a mobility function but 21st century autonomous road vehicles will inevitably come to be seen as providing a mobility service. And the human anthropomorphising tendency ensures that autonomy will be difficult to separate from agency. What should the autonomous road vehicle do when ordered to park in a highly inconvenient but safe place because "I don't want you looking scratched when we get out from the theatre"?

Many examples can be found of the greater psychological complexities of providing a service rather than a function. For example, one study by Apple Inc using their own chatbot technology (Metcalf et al. 2019) found that an AI assistant's likability and trustworthiness improved when the technology mirrored the degree of chattiness of the person using it. The researchers suggested the possible need of adapting the technology's speech characteristics to better mirror the customer.

And a study specifically about autonomous road vehicles (Zhang et al. 2019) asked 443 people to interact with vehicles which were designed to exhibit different perceived personalities. Five personality traits were manipulated:

- extroversion;
- agreeableness;
- conscientiousness;

- emotional stability;
- openness to experience.

And the researchers found that similarities in personality between the vehicle and the participant led to improved opinions of the vehicle's safety and of other vehicle characteristics. The autonomous road vehicle proved more usable and more acceptable when behaving more like the person who was riding in it. Friends often have common characteristics and shared values, and perhaps so will the friendly neighbourhood robots.

While the technical tools for the engineering design of autonomous road vehicles are evolving rapidly, less information and fewer tools are currently available to support their Human Centred Design (Gkatzidou et al. 2021). Human interactions with autonomous road vehicles will involve many issues of perception, cognition, language, emotion, usability, situation awareness, transparency and ethics which did not arise with traditional human-driven road vehicles. For example, at least some new interaction design criteria are likely to be needed for dealing with matters such as:

- anthropomorphic shaping of internal and external components;
- internal and external visual interactions;
- internal and external acoustic interactions;
- internal and external emotional interactions;
- internal and external vehicle behaviours;
- compliance with city ordinances and socially accepted norms;
- accountability and legal liability;
- misuse and self-defence.

The development of the needed design tools may however not be straightforward. One logical difficulty is illustrated by the Naturalness Of Interaction Scale developed by Ramm (see Figure 8.4). Some elements of the scale are obviously appropriate for simple mechanical devices while others are better suited to the evaluation of complex decision-making machines. The difficulty is that the individual elements are not fully compatible in nature or impact. At present, designers have little guidance in relation to the balancing of such aspects of an interaction. Are the simpler and more mechanical aspects more important than those which are more linguistic, psychological or sociological? And does an interaction need to be balanced differently depending on what exactly the autonomous road vehicle is being asked to do by its users at that location at that point in time? Current interaction design techniques do not clarify the trade-offs involved or the factors which merit the most attention.

And to some extent the balance between the more mechanical characteristics and those which are more linguistic, psychological or sociological will depend on the degree of anthropomorphism which the designers choose for the vehicle, which in turn will likely depend on the nature of the service which is provided. Basic point-to-point transport probably does not necessitate sophisticated linguistic, psychological or sociological interactions. On the other hand, autonomous offices, autonomous transport for specialist communities and autonomous entertainment centres may be defined mostly by their linguistic, psychological or sociological interactions.

The history of interaction design suggests that changes in emphasis and the emergence of new techniques have occurred regularly. Each time the nature or complexity of the machines has changed, so too have the interaction design criteria and techniques. And given that our future friendly neighbourhood robots may prove to be more companions than machines, we have probably reached a junction beyond which there will be a fundamentally different emphasis in interaction design.

Conclusions

The word "interaction" refers to physical exchanges or symbolic (communicative) exchanges, or both. It describes situations in which something occurs between two or more entities, with each affecting the other to some degree. This chapter has provided a short introduction to the concept of interaction, to interaction design, to interaction measurements and to interactions with autonomous road vehicles.

The concept of "affordance" was introduced, which is the idea that things can announce their ability to perform certain interactions. Affordances can be of four basic types: physical, functional, sensory or cognitive. This chapter also introduced the concept "control coding" and noted that similar interface controls are usually grouped to some extent according to location, shape, size, mode of operation, labelling or colour.

The concept of "usability" was introduced which refers to the ease of learning, effectiveness in use, and enjoyableness of a product, system or service. Six usability goals were noted: safety, memorability, learnability, utility, effectiveness and efficiency. And the eight golden rules for interfaces were cited: strive for consistency, enable frequent users to use shortcuts, offer informative feedback, design dialogue to yield closure, offer simple error handling, permit easy reversal of actions, support internal locus of control and reduce short-term memory load.

Two important interaction design concerns which impact upon people's goals and happiness were briefly reviewed: frictionless interaction and algorithmic transparency. Frictionless interaction refers to an interaction whose physical, perceptual, cognitive and emotional characteristics are sufficiently intuitive and natural to most people as to provide a negligible step on the way to achieving some goal. Algorithmic transparency refers instead to the person's ability to understand what the machine is doing. It refers to the knowability of the automation's decisions and actions.

The concepts of situation awareness and of user centred design were discussed. And several popular measurement methods which are used by interaction designers were introduced. The NASA Task Load Index (TLX), System Usability Scale (SUS) and Situational Awareness Rating Technique (SART) were presented and the types of interaction which they are best suited to measure were discussed.

The Naturalness Of Interaction Scale was used to illustrate one of the logical difficulties involved in the evaluation of interactions with complex multifunction machines. Some elements of the scale are most appropriate for simple mechanical devices while others are better suited to the evaluation of complex decision-making machines. It was noted that at the present time there is little available guidance to suggest how the different concerns can be balanced when a machine exhibits both.

This chapter noted that only speculations can currently be provided regarding interactions with the future friendly neighbourhood robots. And that the speculations can only be based on facts which are likely to remain somewhat stable going forward in time. Key facts were observed to include the vehicle systems which are likely to need interactions, the operational design domain of the vehicle which prescribes and certifies its usage and the characteristics of the likely services which the vehicle will provide.

And it was noted that the service requirements are likely to lead to many interaction design challenges. Psychologically, 20th century automobiles provided a mobility function but 21st century autonomous road vehicles will inevitably come to be seen as providing a mobility service. Response times, error rates, punctuality, degree of agency and level of politeness all need to be defined. And interaction design targets are also required for the vehicle in relation to the five personality traits of extroversion, agreeableness, conscientiousness, emotional stability and openness to experience.

Important new interaction design criteria are needed for matters such as the anthropomorphic shaping of internal and external components, internal and external visual interactions, internal and external acoustic interactions, internal and external behaviours, internal and external emotional interactions, compliance with city ordinances and socially accepted norms, accountability and legal liability and misuse and self-defence.

Having introduced interaction design and several of the interaction design issues which are of direct relevance to autonomous road vehicles, the next chapter discusses an overarching set of considerations which shape every design decision: ethics.

References

Akamatsu, M. 2019, Handbook Of Automotive Human Factors, CRC Press, Boca Raton, Florida, USA.

Bergstrom, J.R. and Schall, A. eds. 2014, Eye Tracking In User Experience Design, Morgan Kaufman Publishers, Waltham, Massachusetts, USA.

Brooke, J. 1996, SUS: a "quick and dirty" usability scale". In Jordan, P.W., Thomas, B., Weerdmeester, B.A. and McClelland, A.L. (Eds.), Usability Evaluation In Industry (pp. 189–194), Taylor and Francis, London, UK.

Cooper, A. 2004, The Inmates Are Running The Asylum: why high-tech products drive us crazy and how to restore the sanity, Vol. 2, SAMS Publishing, Indianapolis, Indiana, USA.

Cooper, A., Reimann, R., Cronin, D. and Noessel, C. 2014, About Face: the essentials of interaction design, John Wiley & Sons, Indianapolis, Indiana, USA.

Cowen, A.S., Elfenbein, H.A., Laukka, P. and Keltner, D. 2019, Mapping 24 Emotions Conveyed By Brief Human Vocalization, American Psychologist, Vol. 74, No. 6, p. 698.

Crawford, C. 2002, The Art Of Interactive Design: a euphonious and illuminating guide to building successful software, No Starch Press, San Francisco, California, USA.

Diakopoulos, N. and Koliska, M. 2017, Algorithmic Transparency In The News Media, Digital Journalism, Vol. 5, No. 7, pp. 809–828.

Ekman, P. and Friesen, W.V. 2003, Unmasking The Face: a guide to recognizing emotions from facial clues, Malor Books, Cambridge, Massachusetts, USA.

Endsley, M.R. 1988, Situational Awareness Global Assessment Technique (SAGAT), Proceedings Of The IEEE 1988 National Aerospace And Electronics Conference, pp. 789–795.

Endsley, M.R. 1995, Toward A Theory Of Situation Awareness In Dynamic Systems, Human Factors – The Journal of the Human Factors and Ergonomics Society, Vol. 37, No. 1, pp. 32–64.

Endsley, M.R. and Jones, D.G. 2012, Designing For Situation Awareness: an approach to user-centered design, CRC Press, Boca Raton, Florida, USA.

Friedrich, G. and Zanker, M. 2011, A Taxonomy For Generating Explanations In Recommender Systems, AI Magazine, Vol. 32, No. 3, pp. 90–98.

Gibson, J.J. 1966, The Senses Considered As Perceptual Systems, Houghton Mifflin, Boston, USA.

Gkatzidou, V., Giacomin, J. and Skrypchuk, L. 2021, Automotive Human Centred Design Methods, Walter de Gruyter GmbH, Berlin, Germany.

Hart, S.G. 1986, NASA Task Load Index (TLX), Volume 1.0, Paper And Pencil Package, NASA Ames Research Centre, Moffett Field, California, USA.

Hartson, R. 2003, Cognitive, Physical, Sensory, And Functional Affordances In Interaction Design, Behaviour & Information Technology, Vol. 22, No. 5, pp. 315–338.

Hayes, B.E. 1992, Measuring Customer Satisfaction: development and use of questionnaires, ASQC Quality Press, Milwaukee, Wisconsin, USA.

International Organization for Standardization 2002, ISO/TR 16982: Ergonomics Of Human–System Interaction – usability methods supporting human centred design.

Kroemer, K.H. 2017, Fitting The Human: introduction to ergonomics/human factors engineering, CRC Press, Boca Raton, Florida, USA.

Krug, S. 2000, Don't Make Me Think!: a common sense approach to web usability, New Riders, Berkeley, California, USA.

Meiselman, H.L. ed. 2016, Emotion Measurement, Woodhead Publishing, Duxford, UK.

Metcalf, K., Theobald, B.J., Weinberg, G., Lee, R., Jonsson, I.M., Webb, R. and Apostoloff, N. 2019, Mirroring To Build Trust In Digital Assistants, INTERSPEECH 2019, September 15th–19th, Graz, Austria.

Moggridge, B. 2007, Designing Interactions, MIT Press, Cambridge, Massachusetts, USA.

Rader, E., Cotter, K. and Cho, J. 2018, Explanations As Mechanisms For Supporting Algorithmic Transparency, Proceedings Of The 2018 CHI Conference On Human Factors In Computing Systems, Montréal, Canada, April 21st–26th, pp. 1–13.

Ramm, S.A. 2018, Naturalness Framework For Driver–Car Interaction, Doctoral Dissertation, Brunel University London, London, UK.

Rogers, Y., Sharp, H. and Preece, J. 2019, Interaction Design: beyond human–computer interaction, 5th Edition, John Wiley & Sons, Indianapolis, Indiana, USA.

SAE 2018, Taxonomy And Definitions For Terms Related To On-Road Motor Vehicle Automated Driving Systems, Standard No. J3016, SAE International.

Shneiderman, B., Plaisant, C., Cohen, M.S., Jacobs, S., Elmqvist, N. and Diakopoulos, N. 2016, Designing The User Interface: strategies for effective human–computer interaction, Pearson, Harlow, Essex, UK.

Swets, J.A. 1996, Signal Detection Theory And ROC Analysis In Psychology And Diagnostics: collected papers, Lawrence Erlbaum Publishers, Mahwah, New Jersey, USA.

Taylor, R.M. 1990, Situation Awareness Rating Technique (SART): the development of a tool for aircrew systems design. In Situational Awareness In Aerospace Operations, Chapter 3, NATO-AGARD-CP-478, Neuilly sur-Seine, France.

Tourangeau, R., Rips, L.J. and Rasinski, K. 2000, The Psychology Of Survey Response, Cambridge University Press, Cambridge, UK.

Transport Systems Catapult 2017, Taxonomy Of Scenarios For Automated Driving, Transport System Catapult, Milton Keynes, UK.

Tullis, T. and Albert, W. 2013, Measuring The User Experience: collecting, analyzing, and presenting usability metrics, Morgan Kaufman Publishers, Waltham, Massachusetts, USA.

Wachenfeld, W., Winner, H., Gerdes, J.C., Lenz, B., Maurer, M., Beiker, S., Fraedrich, E. and Winkle, T. 2016, Use Cases For Autonomous Driving. In Maurer, M., Gerdes, J.C., Lenz, B. and Winner, H. (Eds.), Autonomous Driving: technical, legal, and social aspects (pp. 9–39), Springer Verlag, Berlin, Germany.

Winfield, A.F., Booth, S., Dennis, L.A., Egawa, T., Hastie, H., Jacobs, N., Muttram, R.I., Olszewska, J.I., Rajabiyazdi, F., Theodorou, A. and Underwood, M.A. 2021, IEEE P7001: a proposed standard on transparency, Frontiers In Robotics And AI, Vol. 8, July, pp. 1–11.

Zhang, Q., Esterwood, C., Yang, J. and Robert, L. 2019, An Automated Vehicle (AV) Like Me? the impact of personality similarities and differences between humans and AVs, AAAI Fall Symposium On Artificial Intelligence For Human–Robot Interaction, November 7th–9th, Westin Arlington Gateway, Arlington, Virginia, USA.

Chapter 9

Ethics

Ethics

Dictionary entries for the word "ethics" usually list at least three concepts:

- the rules of conduct recognised for a particular class of human actions or a particular group or a particular culture;
- a system of moral principles;
- that branch of philosophy dealing with values relating to human conduct, with respect to the rightness and wrongness of certain actions and to the goodness and badness of the motives and ends of such actions.

The word "ethics" therefore refers to either the rules for deciding right or wrong which are used by a group of people in a given situation at a given time, or to the description and study of those rules. Ethics is a part of being human and a key concern of human societies. And its meaning and importance has been the subject of debates spanning back many millennia in human history.

One relatively recent and somewhat scientific view of ethical matters was expressed in 1871 by Charles Darwin (2008) who in *Descent Of Man* wrote "A moral being is one who is capable of comparing his past and future actions or motives and of approving or disapproving of them". Also in *Descent Of Man* he proposed that morality most likely developed due to evolutionary forces, writing that "when two tribes of primeval man, living in the same country, came into competition, if the one tribe included (other circumstances being equal) a greater number of courageous, sympathetic, and faithful members, who were always ready to warn each other of danger, to aid and defend each other, this tribe would without doubt succeed best and conquer the other... A tribe possessing the above qualities in a high degree would spread and be victorious over other tribes; but in the course of time it would, judging from all

DOI: 10.4324/9781003319740-9

past history, be in its turn overcome by some other and still more highly endowed tribe. Thus the social and moral qualities would tend slowly to advance and be diffused throughout the world."

Beyond courage and loyalties, ethics also involves numerous other concepts and concerns which naturally arise in social interactions. While it is not possible to list them all, and difficult even to agree the exact criteria for doing so, Baggini and Fosl (2007) have helpfully identified twenty-one naturally arising concerns which are often discussed in everyday life:

- aesthetics;
- agency;
- authority;
- autonomy;
- care;
- character;
- conscience;
- evolution;
- finitude;
- flourishing;
- harmony;
- interest;
- intuition;
- merit;
- natural law;
- need;
- pain and pleasure;
- revelation;
- rights;
- sympathy;
- tradition and history.

Each provides a basis for connecting thoughts, comparing situations and drawing conclusions about right or wrong. Each can serve as a basis for argumentation and for justifying decisions and actions. And each can lead to a different conclusion about the matter which is under consideration.

Given the range of concerns it is not surprising that attempts were made over the centuries to bring them together. Characteristics and considerations which were found to be common to more than one concern were noted, and the commonalities were expressed at a higher level of abstraction as overarching theories. Introductory textbooks which treat the topic of ethics often contain up to twenty individual moral theories, each involving a set of related ideas which can be deployed when approaching an issue. Each can be considered a viewpoint, backed by criteria, which provides an approach for decision making.

And, to date, no single viewpoint has been found to be more universal in nature or more helpful in practice. Some common human dilemmas have traditionally been analysed using only one or two of the moral theories, but none of the moral theories has been found to be universally applicable or universally preferable. Well-defined and exceptionless moral theories for universal use have not been found. Thus difficult design debates, like difficult political debates, are often the result of the parties having explicitly or implicitly adopted a different moral theory as their reference point.

Given the importance of ethical decisions it is not surprising that the discipline has been developing for millennia and that it currently shows little sign of reaching an end state. The stuff which falls under the heading of "ethics" is as infinite and varied as life itself, thus simple answers to ethical questions are elusive. In most cases the best which can be hoped for is clarity in choosing and communicating the most relevant ethical principles to apply.

Given the complexities the approach adopted here is to introduce a sample of the most frequently cited moral theories so as to illustrate the range of concerns and considerations. For more detailed and complete treatment of moral theories the reader is directed to the several authoritative texts and handbooks which are currently available, such as those of Copp (2005) and Crisp (2013).

Table 9.1 provides a sample consisting of seven moral theories which often appear, explicitly or implicitly, in discussions about design. The sample does not cover all the logical grounds for ethical reasoning and does not claim to be specific to autonomous road vehicles, but nevertheless provides an introduction to the complexities involved in ethical decision making. The sample draws heavily from Vaughn (2015) but also contains theories which are expressed in greater detail elsewhere, such as in Baggini and Fosl (2007), Driver (2007), Gensler (2017) and Warburton (2004).

It is worth noting that some of the theories focus on intentions while others are more consequentialist in nature and are thus more interested in the resulting action. Further, some of the theories are centred on the individual involved, while others attempt to weigh the impacts of the decision or action across all interested parties or even across society globally. Initial choices made in favour of one or the other of the theories lead to potential disagreements with those who had anchored their

TABLE 9.1 Sample of often discussed moral theories.

Theory	Assumptions	Criticisms
Ethical Egoism	The right action is the one which advances one's own best interests. The interests of others are only relevant if helpful to promoting one's own good.	Can be inconsistent since short-term and long-term interests are often in conflict between themselves. Seems to violate a relatively universal component of most ethical theories, that of moral impartiality, i.e. the desire to treat equals equally.
Utilitarianism	The right action is the one which achieves the best balance between happiness and unhappiness, across all of the people who are involved.	It is very difficult to compare the happiness, or the unhappiness, of different people. It is often difficult to decide which exact effects should be considered when evaluating the resulting happiness or unhappiness. It is often the case that the effects of an action differ depending on the time frame considered. Something which leads to happiness in the short term can lead to unhappiness in the long term.
Natural Law Theory	The right action is the one which is consistent with nature, i.e. with its revealed characteristics, purposes and goals.	The approach is absolutist since it has no mechanism for addressing deviations from historical nature. The position that certain acts are always wrong seems impractical and often inconsistent with respect to common sense. The approach fails to provide guidance if the position is accepted that there is no unique, definitive and knowable human nature.
Virtue Ethics	The right act is the one which produces happiness and flourishing due to increasing natural virtues and progressing towards a natural purpose.	It is not always obvious which patterns of a person's feelings, desires and behaviours should be considered virtues. It is not obvious how to deal with situations where the requirements of one virtue of the individual are in conflict with those of another virtue of the individual.

Kantian Ethics	The right action is the one which is performed out of a sense of duty, rather than because of contextual factors such as the effects of the act. Right actions are performed based on good will and duty for duty's sake.	The approach seems incomplete or even empty due to clarifying the reason why a person should act, but not the content of the act. It is not clear in most cases what the duty should be, or why it is a duty. The approach does not take into account the consequences of the chosen act; thus harm can be caused by a right response to a wrong person or wrong situation.
Social Contract Theory	The right act is the one which is consistent with the implicit or explicit social contract in which self-interested and rational people agree the behaviours which ensure their safety, peace and prosperity.	Historically, few people have explicitly consented to a social contract. And it is difficult to suggest that implicit consent occurs simply from living and prospering in a society which is characterised by accepted norms and rules. Contractarian ethics cannot provide a basis for several popular concepts such as that of universal human rights, and thus runs into difficulties with individuals such as children or animals who have only limited abilities to enter into contractual relations.
Feminist Ethics	The right act is the one which corrects how gender operates within our beliefs and practices. The right act supports personal relationships, considers the emotions involved and emphasises care.	Basing decisions on personal feelings, emotions and caring can lead to the breaking of societally accepted moral codes. It appears to be tailorable to each individual and individual situation. If relativist, then it is not a true ethical theory.

reasoning differently. The dichotomies of "intention vs action" and "individual vs societal" have been the subject of repeated analysis from the medieval period onwards, but no simple way yet exists for accommodating all of the perspectives simultaneously.

Historical evidence suggests that one approach for mediating among the possible viewpoints has been attempted from the earliest times in human history: the Golden Rule. The concept of the Golden Rule (Wattles 1996) refers to the principle of treating others as one wishes to be treated by them. Exact wordings of the rule can be either positive in form such as "treat others as you would like others to treat you" or negative in form such as "do not treat others in ways that you would not like to be treated". While focussed more on actions than on thoughts, the Golden Rule provides a first-person perspective which can engage human emotion systems to widen the evaluation beyond the purely analytical.

More recent attempts at articulating global ethical intuitions include the work of Davide Ross (1930) who extended Kantian reasoning by suggesting the concept of prima facie duties. Ross suggested that moral rightness depends on adherence to duties, not on the consequences of actions. Prima facie duties were claimed to be principles which are obvious and which can be prioritised by morally mature individuals. And from among the prima facie duties those which defend or further "virtue" were claimed to be the highest priority while those based on achieving forms of "pleasure" were claimed to be the lowest.

And one of the most recent and influential attempts at mediating among the possible ethical theories is the work of John Rawls (1971) who approached the matter from the perspective of human justice. Rawls suggested that moral principles and priorities should be decided from thought experiments involving what he called an "original position" or "veil of ignorance". The approach involves asking people to express their principles and priorities without knowing what position they would later have in the society, i.e. without knowing what gender, ethnicity, social status or lifestyle would later be theirs. From David Hume onwards it has often been argued (see for example Haidt 2001) that human judgements are the result of innate processes of evolutionary origin which help people to live together and collaborate. Rawls extended the point by arguing that any excesses or unfairness in those human judgements could be minimised if people could not be certain of how the principles and priorities would apply to them personally. While time consuming due to the many needed consultations, Rawls claimed that the principles and priorities which emerge from structured consultations based on a "veil of ignorance" are the best possible mediations and compromises which can be achieved in practice.

Intuitively, global approaches to ethics such as the Golden Rule, Kant's categorical imperative (Kant 1895), Ross' prima facie duties or the principles and priorities defined via Rawls' "veil of ignorance" appear attractive because of their logical rigour. And any global approach which is based on agreed principles and priorities can in theory be handled via deontic logic (permissible, impermissible, obligatory, omissible, optional, non-optional, must, ought, etc.) and could thus be automated to some degree in the manner of Asimov's Three Laws Of Robotics (Asimov 1950). Achieving an algorithmically implementable approach to ethics would in fact prove to be a great simplification, and has thus been a goal of many talented thinkers, past and present.

Achieving a single unified approach to ethical decision making has however not proven straightforward. Numerous weaknesses can be cited in the theoretical underpinnings of the Golden Rule and of the Kantian, Rossian and Rawlsian approaches. And even if most of the theoretical weaknesses could be addressed in some manner, there then remain application issues in relation to the deontic logic.

For example, Powers (2005) and others have noted that judgements can only be the result of consistency tests which compare the given situation with the available deontic options, and that humans deploy a wide range of ideas and experiences when performing such tests. Stating a deontic option linguistically or in some other symbolic manner can prove straightforward, but checking for the existence of that condition among the multitude of characteristics and actions of the real world is not. And as with Asimov's Three Laws Of Robotics (Asimov 1950) it is also not difficult to imagine ways in which consistency tests can fail when attempting to decide between equally important options or between a large number of similar options.

Approaching such matters from the perspective of machine automation, Brundage (2014) identified four key challenges to the algorithmic implementation of ethical principles and priorities:

- insufficient knowledge and/or computational resources for the situation at hand;
- moral dilemmas facing an agent resulting in the sacrificing of something important;
- the morals being modelled by the system are wrong due to insufficient training data, or flawed folk logic or due to flaws in the extrapolation processes;

- loss of understanding or control of the ethical system due to complexity or due to extrapolation of values beyond the current preferences.

At the time of this writing it has proven difficult so far to achieve well-defined and exceptionless moral theories for universal use, and even more difficult to translate them into successful algorithmic implementations. Difficulties arise due to the varied interests and preferred moral theories of the people involved, due to the large number of real-world constraints, due to the lack of certainty about the exact situation and due to the lack of certainty about the exact options which are available. Ethical decision making is thus a complex and time-consuming task with few obvious right answers, and little certainty of universality. Nevertheless, ethical reasoning is a fundamental part of design and some simple tools are in fact available today for detecting ethical issues, for discussing them and for addressing them.

Ethics In Design

One of Heidegger's (1927) many observations was that tools can be understood as connections between humans and reality, and that designed tools such as a hammer disappear from a person's consciousness when performing the act for which it was intended. Csikszentmihalyi and Rochberg-Halton (1981) further generalised the point that, in their own words, "the objects that people use, despite their incredible diversity and sometimes contradictory usage, appear to be signs of a blueprint that represent the relation of man to himself, to his fellows, and to the universe".

In various ways and to various degrees, artists, designers, engineers, philosophers, psychologists, sociologists and others have all claimed that artefacts are unavoidably relational in nature. Designed artefacts in particular have been suggested to be beacons for channelling and shaping the interactions (see Figure 9.1) which occur between a person's will and the triad consisting of the person's other thoughts (self), the other people with which she or he interacts (fellows) and the other objects with which she or he interacts (universe).

And, as with any intermediary, it is difficult to imagine a perfectly impartial and valueless role. Viewing technologies from a postphenomenologist perspective Ihde (1995) stated that "there are no neutral technologies, or, positively put, all technologies are non-neutral". He went on to suggest that technologies and designed artefacts transform the nature, range and quality of human experience. Viewing the matter more

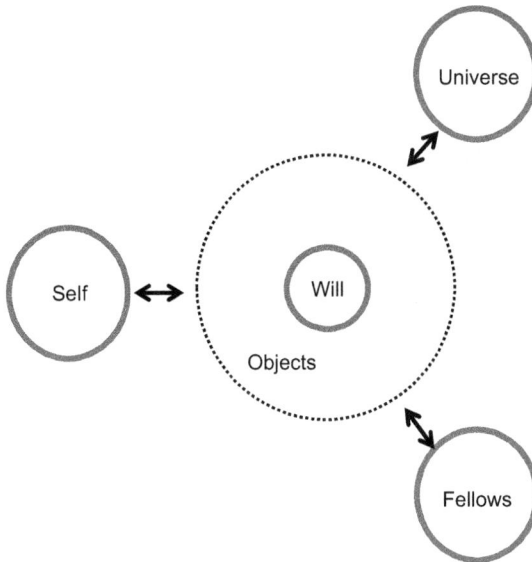

Figure 9.1 The relational role of artefacts.

from a philosophical and sociological perspective, Winner (1986) suggested: "The important question about technology becomes... what kind of world are we making? This suggests that we pay attention not only to the making of physical instruments and processes, although that certainly remains important, but also to the production of psychological, social and political conditions as a part of any significant technical change."

From such positions it is only a short distance further to reach the concept of "ethics by design" which usually refers to the systematic deployment of ethical recommendations and guidelines as part of the design process itself. If all artefacts are intermediaries, and intermediaries cannot be impartial, then design cannot be only about functional considerations. The concept of "ethics by design" expresses the view that the ethical implications of any artefact can be considered during the design process itself, rather than simply releasing the artefact into society and leaving customers and constituencies to experiment.

And some experts argue that even the concept of "ethics by design" does not fully express how designers transform the nature, range and quality of human experience. It has been suggested (Brandes et al. 2013) that designing is actually a form of applied philosophy, i.e. that designers physically experiment the various philosophical principles through their work. On this view, the design process is an applied and pragmatic way of grappling with the questions of what kind of world we live in and of what kind of world we are making.

Typical design issues which are inextricably linked to ethical reasoning include those of accessibility, universality and inclusivity. For proponents of accessible design, universal design or inclusive design, the core objective is to achieve an artefact or an environment which can be used frequently, used comfortably and used economically by as many people as possible (Hanington and Martin 2019). Whether explicitly stated or not, these design approaches involve assumptions which are not distant from those of utilitarian, social contract or even feminist ethics.

Seven general principles of universal design emerged in the 1990s from a project funded by the US Department of Education's National Institute On Disability And Rehabilitation Research (Preiser and Smith 2011). The much referenced principles are:

- Principle 1: Equitable Use.
- Principle 2: Flexibility in Use.
- Principle 3: Simple and Intuitive Use.
- Principle 4: Perceptible Information.
- Principle 5: Tolerance for Error.
- Principle 6: Low Physical Effort.
- Principle 7: Size and Space for Approach and Use.

Principles such as 4, 5, 6 and 7 are aimed squarely at achieving accessibility, i.e. on ensuring full access to services or experiences. Design for accessibility usually focusses on physical, perceptual or cognitive barriers to use, but not necessarily on matters of identify or culture, which can also sometimes affect the outcomes and experiences.

Inclusive design (Keates and Clarkson 2004; Clarkson et al. 2007) is instead a process which usually focusses on users with specific needs, identities and cultures. Often, the user is an extreme user, i.e. someone whose needs and desires deviate substantially from the typical design personas and design scenarios of the sector. Inclusive design attempts to involve diverse individuals throughout the design process and aspires to accommodate all different backgrounds and needs. Universal design principles 1, 2 and 3 are examples of objectives which aim to accommodate diversity of background and of need.

But, as noted by Pullin (2009), many individuals including some who declare some form of disability (Francis and Silvers 2016) can present significant design challenges even when working to the principles of universal design. The wish to support and assist can sometimes prove to be in direct contrast with the need to avoid stigmatising and to pro-vide space for personal expression and identity. For example, a tradi-tional design objective such as simplicity, which is usually considered a positive characteristic, can instead prove disempowering if achieved at the expense of some needed function. Ethically minded objectives and clearly stated universal design principles are unfortunately not always enough to identify the best choices from among the available design options. Doubts can arise due to the varied interests and multiple moral theories of the users and due to the lack of certainty about the exact real-world usage scenario. A degree of judgement is usually an unavoid-able part of the decision making.

And beyond the external-facing characteristics of accessibility, univer-sality and inclusivity there is also a growing focus on the more internal workings of the artefact. Additional areas of "ethics by design" have emerged which include the datasets which are used by the artefact and the data processing which is performed. As artefact complexity has grown and as more decisions have been delegated to it, doubts have arisen about the sourcing of data and about the implicit forms of bias which are present in the data and in the data processing. Data always contains inaccuracies, gaps and biases which lead to consequences of an ethical nature. Ethical risks lurk within datasets.

Popular commercial systems such as Alexa, Cortana and Siri exem-plify the growing concerns because people use their recommender algo-rithms to track down information and to make informed decisions. These systems are important economically, psychologically and socially thus their data gathering, data usage and decisional criteria have come under increasing scrutiny. As complexity has grown and decisional authority has widened, the designers have come under increasing pressure to think through the implications of their choices. Collecting the user's data, displaying it or sharing it can each have unexpected or undesired consequences.

Several organisations have put forward recommendations of good practice in relation to data gathering and data processing. For example, the Organisation for Economic Co-operation and Development (OECD) has recommended a set of guidelines for the protection of privacy and transborder flows of personal data (OECD 2013) which is summarised in Table 9.2. And the European Union has issued its General Data Protection Regulation (EU 2016/679 L119) whose seven core principles are summarised in Table 9.3.

The OECD and EU guidelines describe ethical concerns associated with data collection, data storage and data usage. The guidelines empha-sise the importance of seeking to collect only minimal needed amounts

TABLE 9.2 OECD guidelines on the protection of privacy and transborder flows of personal data (adapted from OECD 2013).

Collection Limitation Principle	There should be limits to the collection of personal data and any such data should be obtained by lawful and fair means and, where appropriate, with the knowledge or consent of the data subject.
Data Quality Principle	Personal data should be relevant to the purposes for which they are to be used, and, to the extent necessary for those purposes, should be accurate, complete and kept up-to-date.
Purpose Specification Principle	The purposes for which personal data are collected should be specified not later than at the time of data collection and the subsequent use limited to the fulfilment of those purposes or such others as are not incompatible with those purposes and as are specified on each occasion of change of purpose.
Use Limitation Principle	Personal data should not be disclosed, made available or otherwise used for purposes other than those specified in accordance with Paragraph 9 except: a) with the consent of the data subject; or b) by the authority of law.
Security Safeguards Principle	Personal data should be protected by reasonable security safeguards against such risks as loss or unauthorised access, destruction, use, modification or disclosure of data.
Openness Principle	There should be a general policy of openness about developments, practices and policies with respect to personal data. Means should be readily available of establishing the existence and nature of personal data, and the main purposes of their use, as well as the identity and usual residence of the data controller.
Individual Participation Principle	Individuals should have the right: a) to obtain from a data controller, or otherwise, confirmation of whether or not the data controller has data relating to them;

TABLE 9.2 Cont.

	b) to have communicated to them, data relating to them
	  i. within a reasonable time;
	  ii. at a charge, if any, that is not excessive;
	  iii. in a reasonable manner; and
	  iv. in a form that is readily intelligible to them;
	c) to be given reasons if a request made under subparagraphs (a) and (b) is denied, and to be able to challenge such denial; and
	d) to challenge data relating to them and, if the challenge is successful, to have the data erased, rectified, completed or amended.

Accountability Principle	A data controller should be accountable for complying with measures which give effect to the principles stated above.

TABLE 9.3 Seven core principles of the EU General Data Protection Regulation (GDPR) (adapted from EU 2016/679 L119).

Lawfulness, Fairness and Transparency Principle	Processing must be lawful, fair, and transparent to the data subject.
Purpose Limitation Principle	You must process data for the legitimate purposes specified explicitly to the data subject when you collected it.
Data Minimisation Principle	You should collect and process only as much data as absolutely necessary for the purposes specified.
Accuracy Principle	You must keep personal data accurate and up to date.
Storage Limitation Principle	You may only store personally identifying data for as long as necessary for the specified purpose.
Integrity and Confidentiality Principle	Processing must be done in such a way as to ensure appropriate security, integrity, and confidentiality (e.g. by using encryption).
Accountability Principle	The data controller is responsible for being able to demonstrate GDPR compliance with all of these principles.

of personal data, of seeking consent for that collection and of seeking to protect against undisclosed or unauthorised use of the information. Explicitly or implicitly, the guidelines assume the ownership of personal data on the part of the human user, seek to minimise the transfer of data to third parties and seek to avoid the ownership by third parties of the information which emerges from analysis of the data. While acknowledging the need for the artefact to know some information about the human user if it is to perform its function, the guidelines describe areas where problems can arise if too much data is collected or if the data is shared or utilised beyond what the human user expected.

And beyond the gathering and use of the data there is also a growing awareness of the ethical risks associated with the selection and tuning of the processing algorithms involved. There are no neutral technologies. The choice of algorithm and the choice of parameter settings can lead to consequences of an ethical nature. Algorithms can display biases in function due to a priori implicit assumptions or due to structures which are present in their underlying mathematics. And such biases can lead to consequences of an ethical nature.

Kraemer et al. (2011) have in fact argued that most algorithms are value-laden and thus implicitly implement rationale or criteria of an ethical nature. The simple balancing of risks, such as those of false negatives and false positives, constitutes a design decision of ethical relevance. Even a design decision as simple as the setting of a numerical threshold value can lead to ethical consequences, since it can act as a divide between domains or between groups of people. Ethical risks lurk within.

One increasingly popular framework which can help when evaluating the ethical risks of an artefact's datasets and algorithms is the set of four general principles (FATE) described by Rogers et al. (2019):

- Fairness: impartial and just treatment or behaviour without favouritism or discrimination;
- Accountability: ability of an automated system to explain how its decisions were made based on the data which was provided by the person or persons;
- Transparency: ability of the automated system to make its decisions visible and to explain how they were made;
- Explainability: ability of the automated system to interact with the person or persons in simple terms which are easy to understand and intuitive for humans.

The FATE principles operate at a somewhat high level of abstraction and are intended as general areas of concern rather than detailed design criteria. Nevertheless, they provide a logical basis for discussing design choices and have been claimed to be fundamental towards the societal adoption of automated systems. Deployment of the FATE criteria at meetings and workshops supports the collection of ethically relevant information, formalises ethical decision making and assists the recording of ethically relevant information which may later prove essential when responding to enquiries, public reactions or litigation.

Extending and expanding upon approaches such as FATE the EU has recently published a set of seven ethical principles which should be adhered to during the development and use of artificial intelligence systems (Floridi 2019):

- human agency and oversight: AI systems should enable equitable societies by supporting human agency and fundamental rights, and not decrease, limit or misguide human autonomy;
- robustness and safety: trustworthy AI requires algorithms to be secure, reliable and robust enough to deal with errors or inconsistencies during all life cycle phases of AI systems;
- privacy and data governance: citizens should have full control over their own data, while data concerning them will not be used to harm or discriminate against them;
- transparency: the traceability of AI systems should be ensured;
- diversity, non-discrimination and fairness: AI systems should consider the whole range of human abilities, skills and requirements, and ensure accessibility;
- societal and environmental well-being: AI systems should be used to enhance positive social change and enhance sustainability and ecological responsibility;
- accountability: mechanisms should be put in place to ensure responsibility and accountability for AI systems and their outcomes.

The seven ethical principles for the development and use of artificial intelligence provide checks which can help to establish the ethical appropriateness, lawfulness and robustness of new developments in the field of artificial intelligence. Designers of complex automated artefacts such as autonomous road vehicles will find the principles useful as points of reference when selecting vehicle characteristics and when choosing between the many possible forms of human interaction. And the articulation of ecological and environmental concerns ensures that the human centred logic also extends beyond immediate short-term goals to the protection of those conditions which are needed to support human life.

And closely related to the matters covered by both the FATE and the EU principles is the issue of trust. Trust has always been an important issue with machines due to the safety, efficiency and economic impacts of malfunction. A machine which cannot be trusted to perform its core function in a reliable manner can quickly become a liability. Features or functions which reveal components or which communicate their momentary operating state are routinely used in machine design to facilitate trust.

The issue of trust has been particularly felt with road vehicles due to the nature of the transport function performed. Malfunction of driving-related components can make it impossible to travel, or at least impossible to arrive on time. And complete failure of a driving-related component can have important safety implications or even be the direct cause of an accident. Access openings, fluid level indicators and dashboard displays have all been a routine part of automotive design in the past and have all helped to stimulate trust in traditional human-driven road vehicles.

While trust has always been a concern of automotive designers, the issue has assumed greater priority with the arrival of the many new forms of automation. As the machine complexity has grown and the machine's decisional authority has widened, the human ability to monitor its function has been stretched and strained. And once the complexity exceeds the human ability to accurately predict the next machine operating state, human judgements can only be based on past operating states. Where human prediction is no longer possible or no longer reliable, the machine can only be judged based on its past "behaviour". There is thus a point beyond which the human judgements must inevitably pass from the realm of "prediction" to that of "trust".

The complex multifunction machines of today are designed to perform many sensing, comparing and deciding tasks which had traditionally fallen within the realm of human activity. Interactions with such machines are noticeably different from those with the machines of the past, and usually involve a higher level of cognitive and emotional complexity. And, as noted in the previous chapter of this book which discussed anthropomorphism, the characteristics and behaviours of such forms of automation can stimulate the human anthropomorphising tendency.

In many cases the interactions with the complex multifunction machines of today are in fact similar to what occurs when interacting with other humans. Think for example of the automotive navigator or of the many online services which involve interactions with chatbots. Prediction of exactly what item of information will be requested next, or of what warning might occur next, is not always straightforward for even the habitual users of the machines. With such complex multifunction machines the establishment of trust would be expected to be a decisive factor in their acceptance and use (Yuen et al. 2020).

Approaching the issue of trust from a psychological and managerial perspective, McKnight and Chervany (1996) claimed that "trust is built or destroyed through iterative reciprocal interaction" and that "the initial period of the relationship is crucial". They identified six forms of trust which can occur in the case of humans, but which are likely to also occur in the case of human interactions with automation:

- trusting beliefs;
- trusting intention;
- trusting behaviour;
- system trust;
- dispositional trust;
- situational decision to trust.

McKnight and Chervany have claimed that the trusting of beliefs is the most influential of the forms, because beliefs underlie intentions, which in turn underlie behaviour. Human values and beliefs form a web of connectivities within which any given intention or any given behaviour will usually reside. Clearly communicating the design principles which an automation reflects and the operating procedures which it adheres to would thus seem to be essential. Such "beliefs" establish the space within which the intentions and behaviours will lie.

And the other identified forms of trust would also seem to be influenceable by the designers. For example, system trust involves declared protections and formally communicated procedures which help people to feel more secure within the environment. Opportunities for the automation to communicate such items of information can be easily designed into its hardware and software systems, along with simple and obvious channels which people can use to interrogate the operational state and forward planning of the machine.

Since trust is the basis for most interactions between humans it would seem logical that it will also prove to be the basis for most interactions between humans and complex multifunction machines. It seems likely that as our ability to instantaneously understand what they are doing diminishes, we will base more and more of our interactions with them on trust. And, as with humans, failures on the part of the complex multifunction machine to live up to that trust will have important and long-lasting consequences, as any brand manager or salesperson knows.

Ethics Of Autonomous Road Vehicles

A large number of physical, perceptual and cognitive driving constraints have been clarified over the years thus a large body of engineering, ergonomic and legal documentation is now in place to support and shape the automotive design process. Many legally binding standards describe the safety and interaction requirements of road vehicles, typically on a category or type basis. And many manufacturers and suppliers have their own internal company standards which specify physical, perceptual and cognitive design targets to meet. Road vehicles are among the best documented and most tightly regulated industrial products.

A large number of the existing standards and targets can be expected to carry over from the current human-driven road vehicles to the future autonomous road vehicles with little or no need for modification. Much of what was established in relation to the physical substrate and physical dynamics of the road vehicle over the course of more than one hundred years of automotive design history will remain unchanged, or only slightly modified.

However, the increasing delegation of the decision making renders the autonomous road vehicle a much more challenging machine from the point of view of ethics. While existing standards in areas such as crashworthiness, ride comfort, thermal comfort and other physical characteristics may require only minor updating for use with autonomous road vehicles, matters involving information, navigation and behaviours will require many new criteria and standards. In particular, criteria and standards will be required for categorising the decisions which are being taken by the vehicle and for clarifying the ethical implications of the various possible courses of action.

That there are ethical concerns in relation to autonomous road vehicles should not surprise. After all, an autonomous road vehicle is a form of robot and the word robot comes from the Czech language word for "worker" or "forced labour" (Capek 2004). Some concerns about what exactly constitutes the "work", and how "forced" that work might be, seem unavoidable baggage of all forms of robot. And the challenge of clarifying the likely ethical issues is not helped by the fact that there is currently no well-defined and exceptionless moral theory for universal use.

And the needed criteria and standards are not necessarily available for translation from the more mature world of computers and information technologies. Where most computers and information technologies are disembodied, the autonomous road vehicle is instead embodied. And where most computers and information technologies are relatively static, i.e. based mostly at a specific physical location, the autonomous road vehicles are instead intended for traversing spatial geographies. Autonomous road vehicles can affect people through physical presence, appearance, touch and movement (Moon et al. 2021) in ways which are not possible with most computers and information technologies. Embodiment and movement, which are considered typical characteristics of living creatures, add multiple additional interactions with humans and multiple additional ethical concerns.

One approach which stimulated many debates in recent years is what is referred to as "trolley problems" (Foot 1967; Jarvis Thomson 1985; Cathcart 2013). A trolley problem is a thought experiment which consists of a highly simplified scenario where a trolley or tram finds itself in an emergency situation where proceeding down one set of tracks will lead to the death of one or more individuals, while shunting onto a second available set of tracks will lead to the death of one or more different individuals. By changing the number of people on each set of tracks or their characteristics (such as age or gender), or by changing the nature of the action required of the tram driver, various ethical dilemmas can be simulated. It has sometimes been suggested that thought experiments such as the trolley problem can be used to produce heuristics for the ethical design and programming of autonomous road vehicles.

However, the multiplicity of moral views and the inevitable situational uncertainties mean that such debates are more about what society values than about what can actually be done in practice. While helpful as a way of establishing the views and priorities of a given society in a given location at a given point in time, trolley problems offer little ethical guidance for the design of autonomous road vehicles which are characterised by a finite number of sensors, finite computational resources, finite training datasets and a noisy operating environment. Autonomous road vehicles have to make split-second decisions based on data streams which are noisy and occasionally erroneous. Recent studies suggest the limited value of trolley problem thought experiments and one (Nyholm and Smids 2016) even questioned whether the analogy can be applied to autonomous vehicles at all.

At the present point in time there is unfortunately not that much ethical guidance which is directly applicable to autonomous road vehicles. And standards, in particular, are few in number. A review of the international standards (BSI and The Transport Systems Catapult 2017) of possible relevance to the design of autonomous road vehicles noted that existing standards are mostly in four areas:

- connectivity/connected vehicles technology;
- connectivity/connected vehicles applications;
- vehicle situation awareness standards;
- vehicle localisation.

Most of the existing standards therefore address functional matters and most are highly technical in nature. Achieving self-driving which actually works, i.e. which actually gets from place to place in a safe manner, has justifiably been the main focus of attention to date. The majority of the existing standards which are relevant to autonomous road vehicles deal with digital communications, mapping and route finding.

Researchers have however already occasionally delved into the ethical characteristics of autonomous road vehicles. For example, in 2010 a joint EPSRC and AHRC Robotics Retreat brought together experts who discussed the design of robots and of autonomous systems. One output from the event was a list of five recommended rules for the ethical design of robots (Boden et al. 2017):

- Robots are multi-use tools. Robots should not be designed solely or primarily to kill or harm humans, except in the interests of national security.
- Humans, not robots, are responsible agents. Robots should be designed and operated as far as is practicable to comply with existing laws, fundamental rights and freedoms, including privacy.
- Robots are products. They should be designed using processes which assure their safety and security.
- Robots are manufactured artefacts. They should not be designed in a deceptive way to exploit vulnerable users; instead their machine nature should be transparent.
- The person with legal responsibility for a robot should be attributed.

The five robot design rules focus on the design intent and appear applicable to the case of autonomous road vehicles. While expressed at a high level of abstraction the rules nevertheless provide a reference for design discussions and for weighting the impact of different design options. Such rules help to clarify what the robot should be doing for people, and its manner of doing.

Focussing instead more on the physical and psychological frailties of the humans who interact with the robots, Riek and Howard (2014) proposed a set of ethical concerns (see Table 9.4) which extend beyond safety and beyond traditional interaction design. The concerns express physical, perceptual, cognitive and emotional issues which should be evaluated by the designers when selecting forms and functions for the robot.

And a recently developed tool which comes close to being an ethical standard specifically for the design of autonomous road vehicles is British Standard 8611 (2016). It describes the ethical issues, ethical hazards and

TABLE 9.4 Ethical human–robot interaction principles (adapted from Riek and Howard 2014).

Type of Consideration	Specific Principle
Human Dignity Considerations	The emotional needs of humans are always to be respected.
	The human's right to privacy shall always be respected to the greatest extent consistent with reasonable design objectives.
	Human frailty is always to be respected, both physical and psychological.
Design Considerations	Maximal, reasonable transparency in the programming of robotic systems is required.
	Predictability in robotic behaviour is desirable.
	Trustworthy system design principles are required across all aspects of a robot's operation, for both hardware and software design, and for any data processing on or off the platform.
	Real-time status indicators should be provided to users to the greatest extent consistent with reasonable design objectives.
	Obvious opt-out mechanisms (kill switches) are required to the greatest extent consistent with reasonable design objectives.
Legal Considerations	All relevant laws and regulations concerning individuals' rights and protections (e.g., FDA, HIPPA, and FTC) are to be respected.
	A robot's decision paths must be re-constructible for the purposes of litigation and dispute resolution.
	Human informed consent to HRI is to be facilitated to the greatest extent possible consistent with reasonable design objectives.
Social Considerations	Wizard-of-Oz should be employed as judiciously and carefully as possible, and should aim to avoid Turing deceptions.
	The tendency for humans to form attachments to and anthropomorphise robots should be carefully considered during design.
	Humanoid morphology and functionality is permitted only to the extent necessary for the achievement of reasonable design objectives.
	Avoid racist, sexist, and ableist morphologies and behaviours in robot design.

ethical risks associated with the use of robots and provides guidance for risk reduction. It subdivides the ethical concepts into the general issue, the specific hazard and the practical risk involved. Table 9.5, adapted from BS8611, summarises the societal, application-based, commercial/financial and environmental hazards.

The BS8611 issues, hazards and risks were agreed by experts who mixed and matched potentially applicable principles and criteria from across multiple disciplines. In some ways such committee work can be said to approximate the "veil of ignorance" thought experiments described by Rawls (1971) since the issues, hazards and risks were selected over the course of lengthy consultations in which the participants did not know, or attempted to not consider, how they themselves might be affected. It can be argued that in absence of a well-defined and exceptionless universal moral theory the principles and priorities which emerge from such structured consultations are the best possible mediations and compromises which can be achieved in practice.

As with the other principles, rules, guidelines and standards discussed in this chapter, BS8611 operates at a somewhat high level of abstraction and provides general areas of concern rather than detailed design criteria. In fact, some of the BS8611 ethical hazards such as those in relation to employment, equality and environmental awareness are issues which have been associated with robots relatively recently, thus they raise questions which are still far from being fully answerable in practice at the present time.

And as with the other approaches discussed in this chapter the deployment of BS8611 issues and hazards at meetings and workshops supports the collection of ethically relevant information, formalises ethical decision making and assists the recording of ethically relevant information which may later prove essential when responding to enquiries, public reactions or litigation. While not necessarily providing direct answers to many of the most important technical and economic decisions, the BS8611 issues and hazards nevertheless provide a tool in support of the decision-making and communication processes.

Acknowledging the lack of a well-defined and exceptionless universal moral theory which can be applied to all aspects of design, researchers have sought technology-specific, context-specific or application-specific alternatives. Testing ethical propositions for only specific personas, scenarios and interactions leverages the constraints to limit the number of options and implications, simplifying the ethical evaluation task. Such efforts have been driven by commercial imperatives but also by scientific claims (see for example Hauser 2006) about the evolutionary nature and general validity of human ethical intuitions. If human ethical intuitions

TABLE 9.5 Ethical issues, ethical hazards and ethical risks (adapted from BS8611: 2016).

Ethical Issue	Ethical Hazard	Ethical Risk
Societal	Loss of trust (human robot).	Robot no longer used or is misused, abused.
	Deception (intentional or unintentional).	Confusion, unintended (perhaps delayed) consequences, eventual loss of trust.
	Anthropomorphisation.	Misinterpretation.
	Privacy and confidentiality.	Unauthorised access, collection and/or distribution of data, e.g. coming into the public domain or to unauthorised, unwarranted entities.
	Lack of respect for cultural diversity and pluralism.	Loss of trust in the device, embarrassment, shame, offence.
	Robot addiction.	Loss of human capability, dependency, reduction in willingness to engage with others, isolation.
	Employment.	Social dislocation, job replacement, loss of skills, need to reskill.
Application	Misuse.	All in this column.
	Unsuitable divergent use.	Unsuitable or inappropriate outcomes.
	Dehumanisation of humans in the relationship with robots.	Inappropriate control exercised by the robot. Loss of respect for human dignity and human rights.
	Inappropriate "trust" of a human by a robot.	Malign or inadequate human control.
	Self-learning system exceeding its remit.	Inappropriate use of resources.

(continued)

TABLE 9.5 Cont.

Ethical Issue	Ethical Hazard	Ethical Risk
Commercial/ Financial	Approbation of legal responsibility and authority.	Failure to meet fair contract conditions leading to illegality, inappropriate actions, avoidance of responsibilities.
	Employment issues.	Job replacement, job change, unemployment, loss of tax revenue.
	Equality of access.	Propagation of the "digital divide", isolation of minorities, non-compliance with human rights legislation.
	Learning by robots that have some degree of behavioural autonomy.	Robot might develop new or amended action plans, or omit steps in processes, that could have unforeseen consequences for safety and/or quality of outcomes.
	Informed consent.	Unaware operators causing accidents, unwanted consequences, unfair and inequitable responsibilities placed upon consenter, inability to respond to situations.
	Informed command.	User is unaware of extent or legality or social acceptance of the tasks given to the robot, consequences of the tasks might ramify in unexpected ways and extents.
Environmental	Environmental awareness (robot and appliances).	Cause concerns about wastage and destruction of the environment. Failure to conform to regulations and/or codes of practice designed to protect the environment resulting in harm to society.
	Environmental awareness (operations and applications).	Execution of non-sustainable actions, harm to local situation, reputational harm.

are reliable, then those ethical intuitions can be directly tested by placing people in the situations of interest. And while only strictly applicable to the conditions tested, such checks nevertheless help to achieve ethically acceptable characteristics and behaviours for the vehicle.

A sizable research literature has in fact already been established for highly specific interactions between humans and autonomous road vehicles. Ethically relevant data has been measured for several internal and external interactions with passengers, pedestrians or other road users. For example, ethically relevant data has been measured for the requesting of the target destination and for the impact upon people of the vehicle's choice of route through the urban landscape. Typically, the studies have investigated the benefits of specific display systems, acoustic announcement systems, virtual reality systems or other such technologies. And also typically, the studies included simple checks of ethical acceptance rather than being detailed studies of the ethical issues involved.

Given the specificity of the research questions which were posed and the limited applicability of the results which were achieved, a review of the contextualised literature was judged to be beyond the scope of this book. The focus of this book has been on considerations which are widely applicable to humans and which are not strictly tied to individual contexts, individual technologies or individual autonomous road vehicles. The review of the highly specific and highly contextualised data is thus a subject for another day.

Conclusions

The word "ethics" refers to rules for deciding right or wrong and to the description or study of those rules. This chapter has provided a short introduction to the concept of ethics, to ethics in design and to the ethics of autonomous road vehicles.

Twenty-one concerns which often serve as the basis for ethical reasoning were introduced: aesthetics, agency, authority, autonomy, care, character, conscience, evolution, finitude, flourishing, harmony, interest, intuition, merit, natural law, need, pain and pleasure, revelation, rights, sympathy, and tradition and history.

And it was noted that introductory textbooks which treat the topic of ethics can contain up to twenty individual moral theories, each involving a set of related ideas which can be deployed when approaching an issue. To date, unfortunately, no single moral theory has been found to be more universal in nature or more helpful in practice. Difficult design debates are thus often the result of the parties involved having explicitly or implicitly adopted a different underlying moral theory.

A sample of seven moral theories was presented to illustrate the range of approaches. The theories were: Ethical Egoism, Utilitarianism, Natural Law Theory, Virtue Ethics, Kantian Ethics, Social Contract Theory and Feminist Ethics. It was noted that some of the theories focus on intentions while others are more consequentialist in nature and are thus more interested in the resulting action. It was also noted that some of the theories are centred on the individual involved, while others attempt to weigh the impacts of the decision or action across all interested parties or even across society globally. Moral theories thus tend to be characterised by the dichotomies of "intention vs action" and "individual vs societal".

This chapter discussed four well-known approaches for mediating among the viewpoints of the different moral theories: Golden Rule, Kant's categorical imperative, Ross' prima facie duties and Rawls' "veil of ignorance". And it was noted that any such global approach can in theory be handled via deontic logic (permissible, impermissible, obligatory, omissible, optional, non-optional, must, ought, etc.).

The difficulty in actually achieving a well-defined and exceptionless moral theory for universal use was however noted, as was the difficulty of translating such a theory into successful machine implementations. Four specific areas of difficulty in machine implementation were listed: insufficient knowledge and/or computational resources, the sacrificing of something important, the modelled morals are wrong due to insufficient training data or flawed folk logic or due to flaws in the extrapolation processes, and loss of understanding or control of the ethical system due to complexity or extrapolation.

Design issues which are inextricably linked to ethical reasoning were suggested to include those of accessibility, universality and inclusivity. And the associated seven general principles of universal design were introduced: Equitable Use, Flexibility in Use, Simple and Intuitive Use, Perceptible Information, Tolerance for Error, Low Physical Effort, and Size and Space for Approach and Use.

Design issues which are inextricably linked to ethical reasoning were also suggested to include the selection and tuning of datasets and algorithms. The OECD guidelines for the protection of privacy and transborder flows of personal data were introduced as were the principles of the EU General Data Protection Regulation (GDPR). And the nature and uses of the popular FATE guidelines of Fairness, Accountability, Transparency and Explainability were discussed.

This chapter also noted that closely associated with the issue of algorithmic transparency is that of trust. A machine which cannot be trusted to perform its core function in a reliable manner can quickly prove to be a liability. Six forms of trust which can impact upon the interactions between people and automation were listed: trusting intention, trusting behaviour, trusting beliefs, system trust, dispositional trust and situational decision to trust.

This chapter suggested that ethical guidelines intended specifically for autonomous road vehicles are not yet available, but that several existing guidelines for robots and for artificial intelligence systems can be considered to be approximately applicable to the case of autonomous road vehicles.

Relevant current guidelines include the EU's seven ethical requirements for artificial intelligence systems: human agency and oversight, robustness and safety, privacy and data governance, transparency, diversity non-discrimination and fairness, societal and environmental well-being, and accountability. And relevant current guidelines also include the five rules for the ethical design of robots: robots are multi-use tools, humans not robots, are responsible agents, robots are products, robots are manufactured artefacts, and the person with legal responsibility for a robot should be attributed.

Finally, a recent tool was discussed which comes close to being an ethical standard specifically for autonomous road vehicles. British Standard 8611 describes the ethical issues, ethical hazards and ethical risks associated with the use of robots and provides general guidance for risk reduction. It was noted that the standard considers a wide range of ethical issues and that some of them have only recently been associated with robots and other forms of automation. As with the other approaches discussed in this chapter, the deployment of BS8611 at meetings and workshops supports the collection of ethically relevant information, formalises ethical decision making and assists the recording of ethically relevant information which may later prove essential when responding to enquiries, public reactions or litigation.

A consideration which was expressed in this chapter is that most current ethical guidelines adopt a substantial level of abstraction, i.e. they provide statements of the principles and objectives involved but lack details about how those ambitions should be achieved in practice. While requirements such as "non-discrimination and fairness" are useful references, they do not suggest how wide the door frame should be or what the driving style should be or what tone of voice the friendly neighbourhood robot should use when asking questions to passengers. The opinion was expressed that, with time, it should prove possible to reduce the level of abstraction from general values to more concrete and prosaic matters such as door aperture sizes, steering angle rates and information system language requirements.

Having introduced the overarching ethical considerations which affect nearly every aspect of autonomous road vehicle design, all that remains in the next chapter is to bring things together and to conclude the narrative of this book.

References

Asimov, I. 1950, I, Robot, Bantam Dell, New York, New York, USA.

Baggini, J. and Fosl, P.S. 2007, The Ethics Toolkit: a compendium of ethical concepts and methods, Blackwell Publishing, Maldin, Massachusetts, USA.

Boden, M., Bryson, J., Caldwell, D., Dautenhahn, K., Edwards, L., Kember, S., Newman, P., Parry, V., Pegman, G., Rodden, T. and Sorrell, T. 2017, Principles Of Robotics: regulating robots in the real world, Connection Science, Vol. 29, No. 2, pp. 124–129.

Brandes, U., Stich, S. and Wender, M. 2013, Design By Use, Birkhäuser, Berlin, Germany.

Brundage, M. 2014, Limitations And Risks Of Machine Ethics, Journal Of Experimental & Theoretical Artificial Intelligence, Vol. 26, No. 3, pp. 355–372.

BSI and The Transport Systems Catapult 2017, Connected And Autonomous Vehicles: a UK standards strategy, March, BSI and The Transport Systems Catapult.

BSI-2016 2016, BS 8611: 2016 Robots And Robotic Devices: guide to the ethical design and application of robots and robotic systems, BSI, London, UK.

Capek, K. 2004, RUR (Rossum's Universal Robots), Penguin, New York, New York, USA.

Cathcart, T. 2013, The Trolley Problem, Or Would You Throw The Fat Guy Off The Bridge?: a philosophical conundrum, Workman Publishing, New York, New York, USA.

Clarkson, P.J., Coleman, R., Hosking, I. and Waller, S. 2007, Inclusive Design Toolkit. Engineering Design Centre, University of Cambridge, Cambridge, UK.

Copp, D. ed. 2005, The Oxford Handbook Of Ethical Theory, Oxford University Press, Oxford, UK.

Crisp, R. ed. 2013, The Oxford Handbook Of The History Of Ethics, Oxford University Press, Oxford, UK.

Csikszentmihalyi, M. and Rochberg-Halton, E. 1981, The Meaning Of Things: domestic symbols and the self, Cambridge University Press, Cambridge, UK.

Darwin, C. 2008, The Descent Of Man, And Selection In Relation To Sex, Princeton University Press, Princeton, New Jersey, USA.

Driver, J. 2007, Ethics: the fundamentals, Blackwell Publishing, Malden, Massachusetts, USA.

European Parliament And Council Of The European Union, Regulation (EU) 2016/679 Of The European Parliament And Of The Council Of 27 April 2016

On The Protection Of Natural Persons With Regard To The Processing Of Personal Data And On The Free Movement Of Such Data, And Repealing Directive 95/46/EC (General Data Protection Regulation), Official Journal, May 4th 2016, L119.

Floridi, L. 2019, Establishing The Rules For Building Trustworthy AI, Nature Machine Intelligence, Vol. 1, No. 6, pp. 261–262.

Foot, P. 1967, The Problem Of Abortion And The Doctrine Of The Double Effect, Oxford Review, No. 5.

Francis, L. and Silvers, A. 2016, Perspectives On The Meaning Of "Disability", AMA Journal Of Ethics, Vol. 18, No. 10, pp. 1025–1033.

Gensler, H.J. 2017, Ethics: a contemporary introduction, Routledge, New York, New York, USA.

Haidt, J. 2001, The Emotional Dog And Its Rational Tail: a social intuitionist approach to moral judgment, Psychological Review, Vol. 108, No. 4, p. 814.

Hanington, B. and Martin, B. 2019, Universal Methods Of Design Expanded And Revised: 125 Ways to research complex problems, develop innovative ideas, and design effective solutions, Rockport Publishers, Beverly, Massachusetts, USA.

Hauser, M. 2006, Moral Minds: how nature designed our universal sense of right and wrong, Ecco HarperCollins Publishers, New York, New York, USA.

Heidegger, M. 1927, Sein Und Zeit, Max Niemeyer Verlag, Tübingen, Germany.

Ihde, D. 1995, Postphenomenology: essays in the postmodern context, Northwestern University Press, Evanston, Illinois, USA.

Jarvis Thomson, J. 1985, The Trolley Problem, Yale Law Journal, Vol. 94, No. 6, p. 5.

Kant, I. 1895, Fundamental Principles Of The Metaphysics Of Ethics, translated by Thomas Kingsmill Abbott, Longmans, Green and Co., London, UK.

Keates, S. and Clarkson, J. 2004, Countering Design Exclusion: an introduction to inclusive design, Springer, London, UK.

Kraemer, F., Van Overveld, K. and Peterson, M. 2011, Is There An Ethics Of Algorithms?, Ethics And Information Technology, Vol. 13, No. 3, pp. 251–260.

McKnight, D.H. and Chervany, N.L. 1996, The Meanings Of Trust, Report MISRC 9604, University of Minnesota MIS Research Center, Minneapolis, Minnesota, USA.

Moon, A., Rismani, S. and Van der Loos, H.M. 2021, Ethics Of Corporeal, Co-Present Robots As Agents Of Influence: a review, Current Robotics Reports, pp. 1–7.

Nyholm, S. and Smids, J. 2016, The Ethics Of Accident-Algorithms For Self-Driving Cars: an applied trolley problem?, Ethical Theory And Moral Practice, Vol. 19, No. 5, pp. 1275–1289.

Organization For Economic Cooperation & Development OECD 2013, OECD guidelines on the protection of privacy and transborder flows of personal data, OECD.

Powers, T. 2005, Deontological Machine Ethics, AAAI Fall Symposium On Machine Ethics, Arlington, Virginia, USA, pp. 79–86.

Preiser, W.F.E. and Smith, K.H. eds. 2011, Universal Design Handbook: second edition, McGraw Hill, New York, New York, USA.

Pullin, G. 2009, Design Meets Disability, MIT Press, Cambridge, Massachusetts, USA.

Rawls, J. 1971, A Theory Of Justice, Belknap Press of Harvard University Press, Cambridge, Massachusetts, USA.

Riek, L. and Howard, D. 2014, A Code Of Ethics For The Human–Robot Interaction Profession, Proceedings Of We Robot, April 4th–5th, University Of Miami School Of Law, Miami, Florida, USA.

Rogers, Y., Sharp, H. and Preece, J. 2019, Interaction Design: beyond human–computer interaction, 5th Edition, John Wiley & Sons, Indianapolis, Indiana, USA.

Ross, W.D. 1930, The Right And The Good, Clarendon Press, Oxford, UK.

Vaughn, L. 2015, Beginning Ethics: an introduction to moral philosophy, W.W. Norton & Company, New York, New York, USA.

Warburton, N. 2004, Philosophy: the basics, Routledge, New York, New York, USA.

Wattles, J. 1996, The Golden Rule, Oxford University Press, New York, New York, USA.

Winner, L. 1986, The Whale And The Reactor, University Of Chicago Press, Chicago, Illinois, USA.

Yuen, K.F., Wong, Y.D., Ma, F. and Wang, X., 2020, The Determinants Of Public Acceptance Of Autonomous Vehicles: an innovation diffusion perspective, Journal Of Cleaner Production, Vol. 270, pp. 1–13.

Chapter 10

Conclusions

Adam Smith (2010) once wrote "that the fitness of any system or machine to produce the end for which it was intended, bestows a certain propriety and beauty upon the whole, and renders the very thought and contemplation of it agreeable, is so very obvious that nobody has overlooked it." According to Smith, propriety and beauty can be found in function.

And if function is so important then it is not so surprising that current design debates about autonomous road vehicles tend to focus on obviously functional matters such as power source, drivetrain, vehicle dynamics, map reading, environment detection and driving control. The engineering options are usually ending up in the limelight. Finding ways of safely moving a person or persons from point A to point B is clearly the current concern.

But the physical properties of the autonomous road vehicles do not provide a complete description of their function. Moving from point A to point B is not the same as being transported from A to B. Humans are profoundly psychological beings thus there are many other issues which impact upon the human experience of being transported beyond simply the physical act of moving. And this book has discussed several design issues which are important when being transported. A series of issues which affect the autonomous road vehicle's ability to meet people's needs, desires and expectations were introduced, as was the philosophy of Human Centred Design (Gkatzidou et al. 2021) which provides a way of discussing such matters and a means for addressing them.

At the start of any design project there are always conceptual issues which, once decided, will affect everything in the process downstream. As illustrated in Figure 10.1 this book has discussed the conceptual issues of anthropomorphism, name, meaning, metaphor, interactions and ethics. Each has an important role to play in shaping the human experience of being transported. Focussing conversations and decision

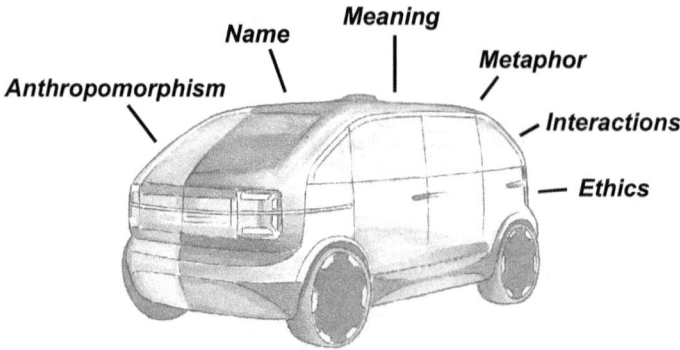

Figure 10.1 Key Human Centred Design issues of autonomous road vehicles.

Source: Henry Leeson

making on these issues does not ensure that the outcome will prove optimal in the eyes of every user, but it helps. Much of the human experience of being transported will be determined by these issues. While representing an unashamedly design-orientated outlook, the topics discussed in this book nevertheless anchor the design process on humans and provide a decision making jury: the people.

Chapter 4 of this book discussed the topic of anthropomorphism. It summarised several key aspects of the phenomenon and discussed how the natural human tendency has traditionally influenced automotive design, and how it may prove even more important in the coming years due to the increased behavioural complexity of the autonomous road vehicles. Chapter 4 implicitly raised the question of whether anthropomorphism is needed to ensure the safety and acceptability of autonomous road vehicles, and that of the appropriate degree and form of manifestation. The benefit or detriment of stimulating the human anthropomorphising tendency will only become fully clear in the coming years as the new vehicles find their way onto our streets. However, designers are already warned that lack of consideration or lack of design targets for this aspect of the autonomous road vehicle may lead to difficulties.

Chapter 5 of this book discussed names and naming. It provided examples of the attribute transferral which occurs with named artefacts, highlighting the possible psychological and sociological implications of the name selection in the case of an autonomous road vehicle. Chapter 5 implicitly raised the question of the suitability of the intended vehicle or service name in relation to the vehicle's characteristics and usage context (geography, travel type, language and culture). As with all designed artefacts, the chosen name can either help or hurt the safety and acceptability of the autonomous road vehicle.

Chapter 6 discussed the meaning of the vehicle or of its mobility service. An overview of the main forms of meaning was provided and examples were cited of how a selected meaning leads directly to many design characteristics. Chapter 6 implicitly raised the question of the role which the autonomous road vehicle will occupy in the lives of its target users. In a world characterised by ever greater complexity and sophistication, artefacts are increasingly designed with value and meaning in mind. And by releasing many core constraints of 19th and 20th century road vehicles, designers will be free to attempt new cabin designs, new packaging arrangements and new service provisions. The material of Chapter 6 was an invitation to designers to explore the possible meanings in depth before putting pen to paper.

Chapter 7 discussed the metaphor which the vehicle or its mobility service alludes to. In a world characterised by ever greater complexity, the artefacts are also increasingly designed with purity and clarity of metaphor in mind. As the functions and choices increase, guiding metaphors become important bridges between the exponentially expanding technical complexity and the finite perceptual, cognitive and emotional resources available to humans. As so many science-fiction stories have prophesised, our technologies are outgrowing us in speed and complexity. And metaphors are among the few simplifying and unifying design tools available for closing the gap.

Chapter 8 introduced the nature of interactions and described some of those which are expected to occur between people and autonomous road vehicles. Several traditional interaction design measures were discussed and the point was made that automation is leading to road vehicles performing more and more functions which were traditionally assigned to humans, and to often performing them in a more human-like manner. More so than the designers of the past, current designers must be skilled at specifying that which is human as much as that which is material and mechanical. Chapter 8 suggested something of a shift from the design of interactions between a person and a thing, to the design of interactions between a person and what behaves much like another person. The more lifelike behaviour of the automated systems is leading to a crescendo of new concerns and considerations. Designers are warned that autonomous road vehicles may involve the psychophysical, psychological and sociological as much, or more, than form and function.

Chapter 9 introduced several well-known moral theories and several of the current frameworks, guidelines and standards which provide ethical guidance of relevance to the design of autonomous road vehicles. Design issues which are inextricably linked to ethical reasoning were suggested to include the traditional concerns of accessibility, universality and inclusivity. Design issues which are inextricably linked to ethical reasoning were also suggested to include the more recent matters of the selection and tuning of datasets and of algorithms. And it was noted that closely associated with the issue of algorithmic transparency is that of trust. It was emphasised that designers will have to be skilled at specifying that which is human as much as that which is material and mechanical. Ethical concerns are immensely important with complex multifunction machines, and in the case of the friendly neighbourhood robots the designers will be crafting what will appear to many people to be more of a new form of life than a new form of transport.

This book discussed the Human Centred Design issues of anthropomorphism, name, meaning, metaphor, interactions and ethics. All of them are claimed to be important and to affect people's thoughts about, and interactions with, autonomous road vehicles. Fully mature design practice in relation to the friendly neighbourhood robots will inevitably require consideration of most of these matters in some detail. Fully mature friendly neighbourhood design practice will only be achieved, however, at some point in the future.

Given that we are not there yet, the question arises of when the knowledge and the skills might be needed. When might design decisions about anthropomorphism, name, meaning, metaphor, interactions or ethics become everyday occurrences and highly critical to success? When might designers need to be at the top of their game in such matters?

Figure 10.2 provides one possible response to such questions by proposing the sequence of introduction of the autonomous road vehicle capabilities. The proposal is based on logical deduction from the known facts about the technologies involved and from discussions with industry experts. Given the difficulty in accurately forecasting technological,

| Full Self Driving On Demand | Business, Entertainment And Healthcare Services | Full Conversational And Emotional Capabilities. |

Time ⟶

Figure 10.2 Proposed order of introduction of new autonomous road vehicle capabilities.

societal and political change, the proposal is ordinal, suggesting only the relative order of arrival rather than the estimated year of arrival. Technical complexity and economic cost have been used to estimate what must come before what.

As suggested in Figure 10.2 it is the author's estimation that safe movement of people from point A to a point B is the current objective of researchers and manufacturers. Overcoming the engineering challenges of fully eliminating the human driver is proving daunting. Achieving full self-driving is thus absorbing much time and effort. And once achieved through a combination of technological excellence and human oversight from control rooms, there are still numerous technical hurdles to overcome in relation to functioning adequately in most road environments in most atmospheric conditions with most humans.

Despite the challenges posed by full self-driving there are nevertheless already proposals for specialist business, entertainment and healthcare services involving autonomous road vehicles. And those services are not being grafted onto, but instead designed into, the autonomous road vehicles. Such proposals involve not simply speciality software but also specialised interfaces, specialty cabin layouts and unique vehicle packaging arrangements. The proposals typically blend the traditional transportation capabilities of road vehicles with specialist services which have usually been provided by humans. Current proposals for mobile offices, entertainment centres and healthcare clinics all suggest that such friendly neighbourhood robots will quickly arrive on the scene once full self-driving on demand is safe and reliable. They are on the near horizon.

And, finally, as also suggested in Figure 10.2 there is the inevitable end point of lifelike friendly neighbourhood robots which are fluent in language, adept at conversation, emotionally aware and emotionally responsive. Eventually, like other future forms of robot, the friendly neighbourhood variety will appear to many people to be more of a new form of life than a new form of transport. And it can be said that designer sensitivity to the natural human desire for social interaction is already leading to some ideas which are rapidly coagulating into feasible autonomous road vehicle concepts. Once around the park, James?

The likely order of appearance of the autonomous road vehicle capabilities suggests that matters such as naming, meaning and metaphor are probably already priority areas for some autonomous road vehicle designs. Public acceptance and commercial success of even small incremental evolutions from current human-driven road vehicles are sensitive to these characteristics. Such matters are probably make-or-break issues for several current commercial propositions.

The likely order of appearance also suggests that sophisticated linguistic abilities, emotional behaviours and awareness of social norms will probably start to prove commercially and societally decisive only in the mid term or the long term. Such matters are not likely to be make-or-break issues for the current commercial propositions, but are likely to be so in the future. The more lifelike and possibly more humanlike the autonomous road vehicles become, the more urgent such matters will be. Once the friendly neighbourhood robots take on more roles within people's lives and provide more meaningful interactions and relationships, any inconsistencies or errors in these capabilities will irritate. Automotive uncanny valley would ensue.

Viewing the situation from where we currently stand it can be noted that there are gaps in the knowledge, criteria, guidelines and skills needed to design the friendly neighbourhood robots from a Human Centred Design point of view. In most of the areas which have been discussed in this book there is a scarcity of guidance at even a high level of abstraction. And none of the frameworks, guidelines or standards provide practical information about the parameter choices involved or the system settings needed to achieve safe and friendly robots.

It is not the case that there is no research literature at all, only instead an issue of focus. Existing resources are largely clustered around contextualised themes such as the possible benefits of specific display systems, specific acoustic announcement systems, specific virtual reality systems or other such technologies. The majority of the studies are highly contextualised and have measured only a small number of predetermined human responses. With few exceptions, the existing research literature is limited in width and applicability. Additional research thus appears necessary before practical frameworks, guidelines and standards of a human centred nature can be formalised.

Examination of the gaps between the research literature and the considerations expressed in this book suggests that further research would prove beneficial. While partial and somewhat subjective, the list below suggests a few areas where new research would be very helpful from specifically a Human Centred Design point of view:

- use of log files for analysis, reporting or as evidence in cases of litigation;
- algorithms for detecting sources of bias in datasets;
- unlearning algorithms for removing sensitive items of information from datasets;
- anonymising technologies for use in passenger recording, situation recording and data exchanges with other vehicles and infrastructure;

- strategies for pickup, parking and energy transfer;
- strategies for communicating the vehicle's intentions to passengers and to other road users via visual and/or auditory descriptions and warnings;
- strategies for dealing with inappropriate vehicle usage by passengers, pedestrians or other road users (ethical misuse, criminal misuse, terrorist misuse, etc.);
- strategies such as driving control by remote operators for use in mechanical breakdown or in emergencies;
- methods and metrics of the stimulation of the human anthropomorphising tendency for autonomous road vehicles;
- methods and metrics of naturalness of interaction for autonomous road vehicles;
- methods and metrics of trust for autonomous road vehicles;
- methods and metrics of robot friendship and for the avoidance of robot dependence;
- strategies for integrating emotion detection systems within the vehicle and outside the vehicle;
- strategies for integrating medical sensors within the vehicle;
- strategies for reducing the motion sickness of passengers;
- medical emergency protocols in support of passengers (contact with emergency services, transport to hospital, etc.);
- emergency protocols in support of pedestrians or other road users;
- physical, perceptual and cognitive inclusivity requirements for specialist communities: children, differently able, non-neurotypical, elderly, etc.;
- strategies for co-designing autonomous road vehicles with associations or with individual members of specialist communities such as elderly or differently abled;
- strategies for personalisation and branding of autonomous road vehicles and services.

With the introduction of the friendly neighbourhood robots we automotive designers are entering a very new field of endeavour. Where previously the relatively static characteristics of form and function were specified, now those same characteristics need to be supplemented with specifications which define behaviour, define the signalling of intentions and enable natural language or other complex forms of communication. Emotional interactions with humans may also prove beneficial, perhaps necessary, thus also part of the design remit and design brief.

Designing a road vehicle which can accommodate a child safety seat is a relatively straightforward matter involving appropriately sized mechanical fixtures, adequate ingress and egress arrangements and crash safety testing. Much more will instead be required in the case of a friendly neighbourhood robot which must transport a child safely and enjoyably, perhaps without the assistance of a parent or guardian. Such a design challenge is both new and daunting. The designer might leverage some forms and some functions from traditional automotive practice, but most of the interactions, communications and emotional exchanges are a brave new world.

In the Autonomous Era the crafting will inevitably prove to be less about inanimate objects and more about products, systems and services which appear to be forms of life. A time traveller from the distant past, born into a world where travel was by foot or beast, would be confused by 20th century motor vehicles which move at the mere push of an accelerator pedal. If thrust into a world full of friendly neighbourhood robots, however, that same traveller might lose all sense of reality as the strange machines shuffle back and forth without human intervention. Designers need to prepare for the Autonomous Era.

And somewhere over the horizon there is also the matter of robot rights (see Gunkel 2018 for a review). When designing the friendly neighbourhood robots many of their characteristics will be constrained by their moral and legal position within human society. As a current area of intense philosophical, sociological and legal debate, there is growing recognition that some legal rights will be assigned to friendly neighbourhood robots as had happened with humans, then animals, and most recently with some elements of the natural environment. Whether due to the machines reaching a degree of consciousness (Will Theory) or due instead to the effects on human society of how we treat them (Interest Theory), some moral and legal recognition of some robot rights seems imminent. If for no other reason than because granting our friendly neighbourhood robots some rights is to err on the side of caution (Neely 2014). It is impossible to predict the unintended consequences of robot mistreatment, robot subjugation or robot slavery. Thus robot rights may be a necessary instrument not for protecting the robots, but for protecting the humans.

The history of design suggests that changes in emphasis have occurred regularly. Each time the nature or complexity of the machines has changed, so too have the design criteria and design techniques. In the Autonomous Era our future friendly neighbourhood robots may be more companions than machines. We have thus probably reached a junction beyond which there will be a fundamentally different manner of designing. One fully centred on, and revolving around, humans.

References

Gkatzidou, V., Giacomin, J. and Skrypchuk, L. 2021, Automotive Human Centred Design Methods, Walter de Gruyter GmbH, Berlin, Germany.

Gunkel D.J. 2018, Robot Rights, MIT Press, Cambridge, Massachusetts, USA.

Neely, E.L. 2014, Machines And The Moral Community, Philosophy & Technology, Vol. 27, No. 1, pp. 97–111.

Smith, A. 2010, The Theory Of Moral Sentiments, Penguin Books, New York, New York, USA.

Index

For Product Safety Concerns and Information please contact our EU
representative GPSR@taylorandfrancis.com
Taylor & Francis Verlag GmbH, Kaufingerstraße 24, 80331 München, Germany

www.ingramcontent.com/pod-product-compliance
Lightning Source LLC
Chambersburg PA
CBHW070328270326
41926CB00017B/3811